Office of the
Deputy Prime Minister

Creating sustainable communities

IPACT OF
WARDS
OUSING

April 2004

Mike Biddulph
Professor Alan Hooper
Professor John Punter
Department of City and Regional Planning, Cardiff University

Office of the Deputy Prime Minister: London

The Office of the Deputy Prime Minister
Eland House
Bressenden Place
London SW1E 5DU
Telephone 020 7944 4400
Web site www.odpm.gov.uk

Further copies of this publication are available from:
RIBA Enterprises Ltd
15 Bonhill Street
London
EC2P 2EA
Tel: 020 7256 7222
Fax: 020 7374 2737
Email: sales@ribabooks.com
or online via the Office of the Deputy Prime Minister's web site.

ISBN 1 85946 156 5

Printed in Great Britain on material containing 75% post-consumer waste and 25% ECF pulp.

April 2004

Ref No: 04PD02093

Advisory Group

For their patient and kind assistance with this project we would like to thank the following people who formed our Advisory Group:

Mark Cousens, Office of the Deputy Prime Minister, Urban Policy

Alex Ely, Commission for Architecture and the Built Environment

Tina Golton, Office of the Deputy Prime Minister, Research Analysis and Evaluation Division

Stephen Mullin, *Housing Design Awards*

Simon Pinnegar, Office of the Deputy Prime Minister, Research Analysis and Evaluation Division

David Scott, DMS Consulting, Gloucester

In addition we would also like to thank all those companies and individuals who generously assisted with this work, in particular by offering information and agreeing to be interviewed.

Contents

Foreword

This report gives all of us involved with the award system some real food for thought and basis for action. I first declare an unfortunate prejudice in introducing this important piece of work – I dislike the word 'housing'. To me it implies the residential monoculture that we have created in my lifetime in the form of housing estates and urban sprawl.

The RIBA's principal mission is to advance the art and understanding of architecture, to improve the built environment and consequently the lives of all of us who live, work, learn and play in it. We believe that good architecture is fundamental to the success of a healthy society and that the award system plays an important part in helping to raise the game. That is why we have been involved with Government in the *Housing Design Awards* since 1947 when the great Nye Bevan introduced them for public housing. Over the years, the RIBA and Government have been joined by the RTPI and NHBC, and the awards have embraced the fast growing private sector and of course the Housing Associations and other agencies which have all but replaced public housing as we knew it.

Do we expect too much from design awards for housing, and the *Housing Design Awards* in particular? There are pointers to all involved in the *Housing Design Awards* – ODPM, RIBA, NHBC and RTPI – about areas we could strengthen. I know from the RIBA's involvement at the consultation stage that the organisers have already responded constructively to some of the criticisms. For 2005 we have increased the vital 'lay' element in the judging and will be engaging much more with users about the performance of shortlisted schemes in the broadest sense.

Furthermore, the Housing Design Awards, which are often wrongly perceived as belonging solely to the RIBA rather than the joint initiative of four partners, will this year, for the first time in my memory, not be announced at Portland Place, but at Housebuilding 2004 at the Business Design Centre. They will also be promoted to a wider public through a more accessible and readily available book, and an improved website and touring exhibition. For the first time there will be the added glamour of an overall winner, aimed at raising the media profile and the public perception of the importance of design in this vital field.

Awards are one element in what needs to be a comprehensive approach to raising standards. The good examples are growing in number and quality and architects are getting more and more involved, but I believe that we should be intolerant of the mediocre and should shame the bad examples, which are all too prevalent. Government has encouragingly strengthened the role of design in the latest planning legislation and through such measures as the establishment of CABE and are taking increasing responsibility for the quality of the built environment. Now it is up to us as architects, planners and developers to take up these opportunities to create better homes and places for people to live in.

George Ferguson
RIBA President

Chapter 1

Introduction

Chapter 1
Introduction

Awards for housing design provide the opportunity to reward, celebrate and encourage the best in residential design. They also provide the opportunity to learn about new forms of living environment and the potential quality of development which housing developers and designers can deliver. Such issues are of paramount importance as the quality of living environments has a significant impact on our quality of life. New planning and design agendas seek to promote an 'urban renaissance' to which design can contribute. Improving the quality of residential environments is central to encouraging people to live in established urban communities, at higher densities, and in environments that support more sustainable patterns of life.

Rewarding, celebrating and encouraging forms of housing that embrace this agenda should therefore be a goal of Government. The potential of design awards to stimulate new ideas and provide positive examples of good design from which others can learn should also be exploited.

What is a housing design award?

An award for housing design is given for either designs or completed schemes where the designs have been commissioned for a range of unspecified sites and by third parties unrelated to the award giving process. This contrasts, for example, with a design competition, where a group of interests encourage designers to submit a range of schemes relating to a particular site or development scenario.

Recognising this potential, the Office of the Deputy Prime Minister (ODPM) currently sponsors one of the principal awards in the housing field – the *Housing Design Awards* – and the particular role the *Housing Design Awards* plays within the wider awards portfolio is a central interest of this research. The importance of good design quality, and the contributory role the *Housing Design Awards* seek to play in encouraging improved practice is emphasised in the forward to the 2002 awards booklet:

> *"If we are to leave a legacy of high quality housing to future generations, and ensure the successful regeneration of our towns and cities, the principles set out in Planning Policy Guidance 3 (PPG3) must underpin the design of any future development. We need to maximise the use of brownfield sites and ensure that any new urban extensions are both compact and sustainable…The award-winning schemes shown in this book demonstrate how this can be achieved."* ***(Anonymous, 2002)***

No previous research has been done to explore the extent to which design awards in general contribute to the wider objectives of Government design policy. Of particular interest in this piece of research is whether house building companies use, or could be encouraged to use, lessons learned through participation in, or from the results of, design awards to develop their product in line with Government aspirations. In other words, do companies look at and learn from award winning exemplars and reconsider their approaches to design and construction, and if not how might the awards be changed to promote good design in the future?

The objective of this research is to consider how the *Housing Design Awards* and other design awards for housing contribute to encouraging better design and in so doing complement Government policy objectives. To do this it has been necessary to find out how particular groups respond to the range of award programmes that run in England, and to look at the range of factors that encourage or discourage interest and participation. In particular, it has been necessary to find where tensions occur between what people want from housing development and award programmes, and to consider a range of options which highlight how an award programme might be designed to be attractive to one group or another.

To meet this objective, a number of more specific aims were formulated to provide a way of analysing both the design awards themselves and the attitudes and interests of the groups that have been targeted in this research:

- to review the principal awards presented in the housing design field and provide an assessment of each scheme's aims and objectives, format and structure, intended audience, perceived status and influence within the house development sector;
- to assess the views of housing developers and professionals including architects and planners towards design awards generally, and the *Housing Design Awards* specifically, to understand how design awards are viewed and how they influence industry practice;
- to consider how housing consumers' decisions to buy or rent a property are influenced by an awareness that a scheme has won a design award, and in particular whether this has encouraged them to live in a scheme which has been built, for example, to a higher density and on brownfield land;
- using the findings from the above, to highlight how interests respond to the key design awards in this sector and to consider how aspects of a design award programme might be designed to be more attractive to those interests, as well as consider the compromises that might be necessary in order to target an award to any particular group;
- to provide recommendations in the form of key options for the future form of the *Housing Design Awards*, and to consider the implications of following any particular approach.

Despite the emphasis in the recommendations on one award programme, this research has had to consider how key interests respond in general to awards for housing design. The results and analysis should therefore be of interest to any organisation, including local authorities, seeking to use an award programme to encourage better design. It is also hoped, that the results will be of more general interest to professionals involved in residential development, such as house builders or architects, reflecting on both why they enter design awards and how they might more positively use award results to develop their practices.

Approach to the study

The empirical work and analysis conducted during this research have been divided into eight parts:

Literature Review

An initial literature review focused upon professional surveys and academic literature which consider questions of preference in the design of both houses and residential schemes, how 'professional' and 'lay' tastes are formed, and how competitions can be organised to reflect such preferences. This sought to explore why particular people respond to different forms of housing, or focus upon different aspects of housing in different ways. Mechanisms available to Government for influencing the form of residential development were then considered to help locate design awards for housing within the broader context. Finally, the current policy context for encouraging and requiring greater attention to housing design was assessed in order to better understand the forms of housing that Government seek to encourage as part of their urban renaissance agenda and also why.

Award Analysis

A search was then undertaken to find relevant English awards for housing design. This was undertaken using the web, professional and trade periodicals, newspapers and discussions with planning, design and development professionals. Initially, all awards related to housing were reviewed, and from this, awards specifically related to housing design in England were determined. In order to provide good representation of recent activity, information was collected relating to the period 1997 – 2002. Documentation from the selected design awards was analysed and a telephone interview completed with either the Chair of the judging panel or another person recommended by the Chair. The interview provided details of the design awards' aims and objectives, format and structure, intended audience, perceived status and influence within the house development sector. Information gathered through the interviews was used to compare both the design awards' characteristics and the types of participation from different categories of house builder in the different programmes.

Whilst collating information about the award programmes, it became clear that this was the first time that anyone had tried to create a coherent picture of them and present it in an accessible and integrated format. To a certain extent, therefore, the organisations running the awards were not prepared for the specific questions that were asked, and sometimes the types of information that could be gathered were not consistent. Nevertheless, the analysis presents a clear, robust impression of award programmes and award winning schemes.

Award Winning Case Studies

In order to assess the views of housing developers, architects and planners to design awards generally, and to consider how housing consumers had been influenced by a scheme's award-winning status, it was decided that 13 award winning schemes would be used as case studies. This was undertaken by selecting house builders whom had been successful in one, or indeed a number of, the selected award programmes. Schemes were also selected as far as possible according to location, market and tenure variation. The developers approached – 5 volume house builders, 2 medium sized house builders, 3 small developers and 3 social housing providers – reflect the range of housing providers, with a slight emphasis on the larger companies. A good variation in terms of context was also sought in order to explore the widest range of housing being developed, without emphasising any particular housing type. As such, it is hoped that the conclusions presented are relevant to a wide audience. A list of interviewees is included as Appendix 3. Details of each case study can be found in Appendix 4.

Interviews with developers and professionals were used in order to discuss whether, and how, the design awards have encouraged greater attention to design and urban renaissance concerns. They were also used to help explore which types of design award the interviewees preferred and why. This made it possible to understand the key tensions, if any, between these groups, and identify the features that motivate them to become involved in some award programmes and not others.

Interviews with residents and investors aimed to find out if their decision to buy their homes had been influenced by the fact that the scheme had won an award. Getting residents and investors to provide comments was sometimes difficult since the majority of schemes included apartments with carefully controlled access arrangements. Indeed, a number of the housing schemes formed gated communities for which access consents would be required. Where possible, interviews were also held with sales and marketing professionals working within the relevant companies in order to determine their view as to how consumers respond to the award winning status.

Gated entrances at Century Court, Cheltenham

Non-Participant Interviews

To better understand why some companies may chose not to get involved in design awards, contact was also made with house builders who appeared to have no track record in terms of winning awards. Given that most of the award schemes only keep records of winners rather than entrants (if any records are kept at all), it was difficult to ascertain from analysis of awards information which companies have not engaged in the awards process. Through contact with non-winners, it was found that most of the larger firms have some record of entry in at least one of the award programmes. As a result twenty smaller firms were targeted. However, these firms typically work in a regional market and develop only a small number of houses in any year and as a result did not feel that they could make a positive contribution to the research.

Wider Analysis of Developers' Output

It can be asserted that developers typically select and put forward only their best schemes for inclusion in award programmes, with other schemes within their wider portfolios perhaps not designed and developed to similar design and quality standards. Development brochures were collected for schemes by the volume companies and studied in order to characterise in more general terms the companies' approach to design. This analysis helped determine whether considered selection and promotion of schemes occurs and this helped supplement discussions in the interviews which sought to determine if experience and lessons learned through award winning schemes were being translated across wider company practice.

Media Interviews

Interviews with representatives of media promoting award winners were conducted in order to gain perspectives as to whether and to what extent there is public interest in award schemes and associated publicity. In particular, these interviews were used to explore how the awards could be given greater or better coverage to bring them to the attention of a wider public. Media contacted included national and regional newspapers and public relations professionals. Attempts to interview television production companies were unfortunately not successful, although a number of the interviewees had good experience of the television media and could provide advice about how they might be made more interested in housing design awards. A list of people and organisations that participated in these interviews is included in Appendix 3.

Choices in the Design of the Housing Design Awards

This research has made it possible to determine key themes that characterise the choices available in formulating housing design awards. It can be suggested that the ideal award programme would be one in which all of the key interests would see genuine benefits and would be motivated to become involved. Equally, the process of being involved in such an award would impact positively on other areas of design and development work. The research highlights, however, that differences exist between interests, and that certain approaches to formulating a design award programme tend to benefit one set of interests over another. The analysis draws out these points of difference and relates them to the features of the award programmes to determine the choices that seem to exist. This allows anyone designing such a programme to be able to target the award towards any given set of interests, but at the same time be aware that other interests may be less engaged.

Recommendations for the *Housing Design Awards*

The report concludes by suggesting a series of options for how the *Housing Design Awards* might evolve. Three options are suggested. Each points out both the benefits and the potential disadvantages of each approach, whilst links between the options and the themes explored through the research are also highlighted.

The rest of this report

Chapter Two draws upon the literature review to consider the issues which shape our preferences in housing design, and how these preferences can influence approaches taken and reactions to award programmes. **Chapter Three** considers the role and scope of the variety of mechanisms available to Government in either controlling, guiding or encouraging better practice in housing design and layout – awards being an example of mechanisms for encouragement. Details of awards available for housing design in England are then introduced and their characteristics compared in **Chapter Four**. **Chapter Five** provides an analysis of how actors respond to both the varying nature of these design awards, in the context of their wider appreciation of design quality and understanding of the broader housing design agenda. Building upon this, **Chapter Six** compares the sometimes contradicting interests of these actors with the characteristics of the range of awards for housing design as a means of illustrating how any award might be altered to make it more or less attractive to one group or another. In **Chapter Seven** final conclusions and recommendations are presented in the form of a series of options for the future of the *Housing Design Awards* in particular.

Chapter 2

Judging Housing Quality

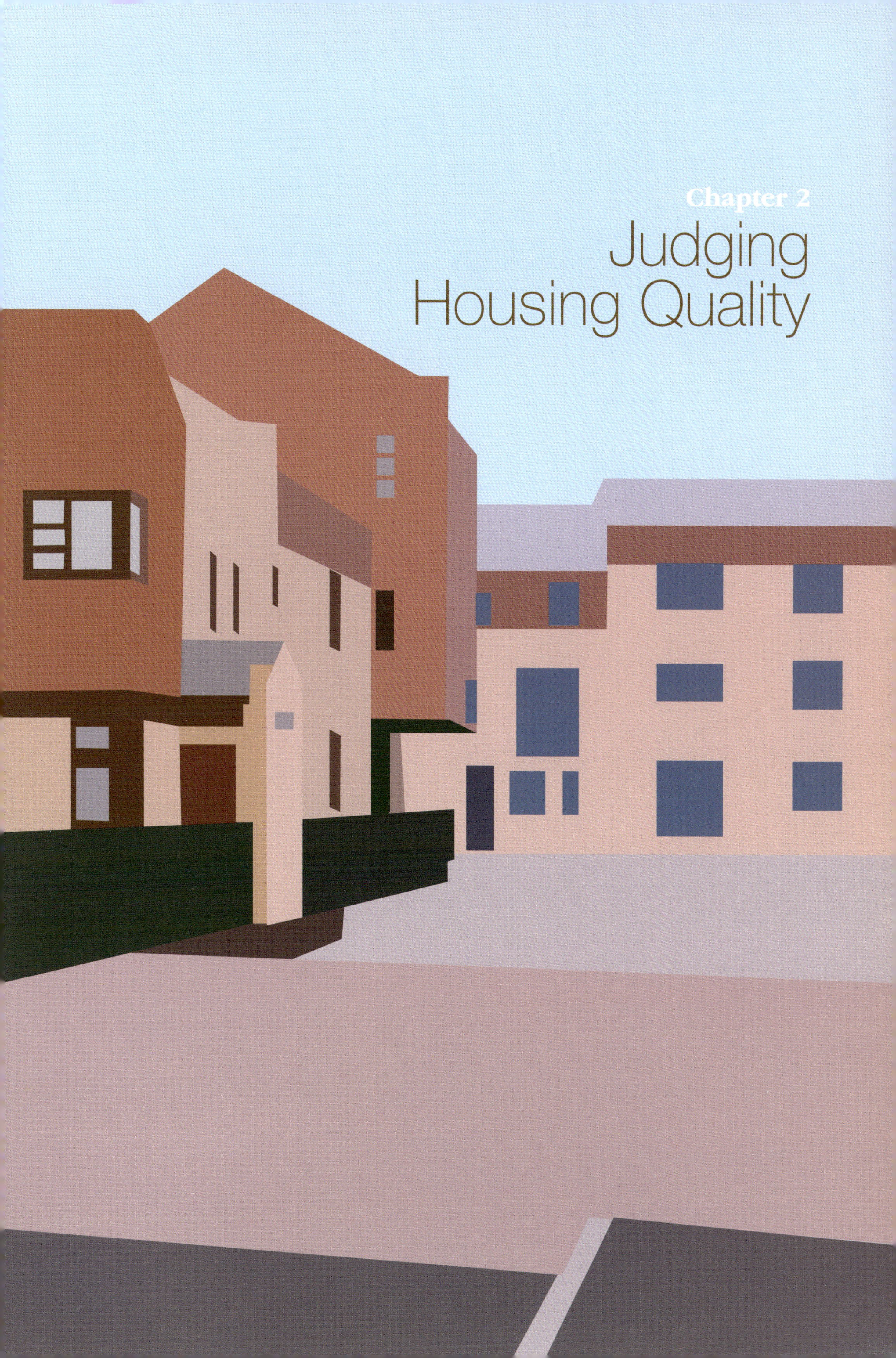

Chapter 2
Judging Housing Quality

Only relatively recently have factors that shape our preferences in the design of both houses and residential schemes been explored within professional surveys and academic literature. The latter has highlighted the significance of perspectives in the judgement of design quality, and therefore preference. In particular, debate has focused on a tension between consumer and public interests in residential layout on the one hand, and between professional and lay tastes in residential design on the other. These perspectives are considered in this chapter in order to provide a framework to help explain the responses found through the research's case study work to design awards for housing. Studies on the process and features of award giving in particular also provide pointers as to how such design awards might be formulated to accommodate different perspectives.

Moving away from definitions of 'good taste'

The question of taste is inevitably a highly contentious matter. Discussions about design quality have traditionally been rooted in attempts to define visual beauty in terms of the physical qualities of a building. Such an approach depends on the assertion that certain formal qualities or styles are innately beautiful and, as a result, are universally regarded as being in 'good taste'. The Royal Fine Art Commission's treatise on *What makes a good building* continues this tradition, espousing a set of criteria as objective values around which consensus could be reached rather than considering the issue of stylistic preference as a matter of taste which as such is largely subjective (RFAC, 1994). It is a laudable aim to try to distance debate about good design from a 'battle of the styles'. However, it can be argued that this fails to acknowledge the broad range of meanings attached to buildings and the environment, preferring to trust in the consensus achieved by '*experienced designers and informed laymen*' who could recognise quality when they saw it and be objective about it (p.80).

Narrow formal conceptions of environmental aesthetics have been replaced by broader notions which focus more on the symbolic, cognitive and non-sensory meanings of design. These newer perspectives recognise that environmental meanings are socially constructed and transmitted (Bourassa, 1991; Punter, 1994) and that rather than being highly object oriented and idiosyncratic, they signify values that define individual and group identities (Alcock, 1993). Researchers, in particular, have explored the meanings that people attach to buildings and places and how perceptions and evaluations of the built environment vary between different individuals and groups. Differences have been investigated in terms of class, gender (especially important in housing) and age, but the most clear-cut distinctions have been drawn between those people trained in environmental design, particularly architects, and those who have had no training (Devlin, 1990). It is useful to note, however, that professional designers are socialised into a culturally prescribed role with a distinctive self-image, culture, language and professional ethos (Salaman, 1974; Uzzell and Lewand, 1989; Cuff, 1991; Bentley, 1999).

Similarly, the planning profession shares some elements of this ethos overlaid with "*an amalgam of paternalism, environmentalism, aesthetics and social determination*" (Knox and Allen, 1981; Hubbard, 1994).

This understanding is thought to provide a context for understanding how judgements about housing preference are formed, and this has been explored through the use of case studies and discussions with the various actors involved in the research. Rather than seek consensus on matters of '*good taste*' this research recognises that preferences are based to some extent on experience, and that these preferences are themselves shaped by social affiliations. This position reflects that there is no basis upon which objective criteria for good residential design might be based, although preferences will be clearly apparent.

Professional versus lay tastes in development control and design review

Studies of the outcomes of British planning development control and their relationship to public design preferences are rare (Hubbard 1994). By comparison, US researchers have been particularly keen to examine design review practices in US planning to test whether the efforts of planners and review boards in evaluating designs are appreciated by the general public and thereby serve the public interest. In the US zoning-based system, where there are clear development rights, there is much more scepticism about design regulations and the exercise of discretion in matters of design (Nasar and Grannis, 1999). There is a belief that many such interventions are arbitrary, capricious and vague, and even counter-productive. Research has explored whether design codes and processes deliver their intended benefits, whether practical consensus can be achieved about the desirability of particular decision outcomes, and how to identify specific and explicit criteria to guide the regulatory process (see Stamps and Nasar, 1997; Nasar and Grannis, 1999). A consistent finding of this work is the difference in views between design professionals on the one hand and local residents and the general public on the other when evaluating external appearance and design quality (Brewer, 1988; Nasar, 1994 and 1989).

Linda Groat (1994: 164) has done extensive work exploring lay and professional approaches to the question of contextual fit. She notes that:

> *"...whereas the issue of visual continuity is only one of several possible goals in architects' contextual design strategies, it appears to be the primary goal for most non architects."*

Groat's analysis emphasises that architects and architectural critics "*undervalue certain popular codes in interpreting the environment*", and particularly the desire for visual continuity. She draws comfort, however, from the fact that design review commissions with wider professional and lay representations, "*are not likely to let their particular architectural preferences substantially influence their decisions*" about acceptable forms of development (Groat, 1994: 164).

In a British context (Central Birmingham in the early 1990s) Hubbard notes that planners put:

> *"...more emphasis on whether a building fitted into its surroundings, its use of appropriate materials and technology and its design approach, whilst the public appeared to prefer developments which incorporated decorative architectural motifs and details...because of the contribution they made to people's sense of stability and identity."* ***(Hubbard, 1994: 282-3)***

The latter point does not refer to housing developments, but would seem to be particularly relevant. Hubbard concludes that:

> *"...planners need to develop a more considered awareness of local needs and wants, as well as interrogating and articulating their own aesthetic preferences to close the gap between lay and professional tastes."* ***(ibid: 283)***

The next section highlights more specifically research that has explored what types of housing people prefer, and considers if there are fundamental contradictions between the preferences that groups express.

Consumer, public and professional satisfaction with housing design

There is little evidence to suggest widespread consumer dissatisfaction with the quality of new housing in Britain, although recent research does point to some particular concerns regarding design, the lack of choice in housing and a desire for more modern forms (Popular Housing Forum 1998; The Housing Forum, 2001; New Homes Marketing Board, 2002). Most industry-derived consumer research reveals strong preferences for bungalows (MORI poll for CABE, 2002) and detached housing in low density estates in relatively homogeneous neighbourhoods in suburban or village locations, though this is clearly very much the perception of nuclear families with children (Mulholland Research Associates, 1995).

Rather, much of the dissatisfaction with contemporary private housing has been expressed by:

- the planning profession and planning committees who have to approve all new development;
- local residents whose amenities will be affected by it;
- architects who rarely get to design it; and
- a range of environmentalists who object to its impacts upon the countryside and, in the last decade, its sustainability implications.

A recognition of the importance to reflect public interest in planning decisions is seen in the proliferation of design guidance used by planning authorities to influence the form of housing schemes. Initially, such guidance was seen to be preoccupied with the visual quality of the external environment (see for example Essex County Council, 1973). As a result, the social perspective of how a dwelling is lived in and the key design qualities that dictate resident satisfaction were largely ignored (see for example Marcus and Sarkissian, 1986). Circular 22/80, which emanated from a well-orchestrated developer campaign against such residential design guidance, stated that:

Most industry-derived consumer research reveals strong preferences for bungalows and detached housing in low density estates

"Developers should not be compelled to conform to the fashion of the moment at the expense of individuality, originality or traditional styles. Nor should they be asked to adopt designs which are unpopular with their customers or clients." ***(DoE,1980, para. 19)***

However, it also concluded that local authorities should ensure that developments were not:

"out of scale or character with their surroundings." ***(DoE, 1980, para. 20)***

The contradiction within these statements is clearly apparent, and reveals two important and recurring features within the design quality debate in Britain. Firstly, the "battle of the styles" has occupied a particularly important place in the debate, regularly fanned by both traditional and modern architects, local amenity societies and residents' groups. Secondly, the concern that development should acknowledge its context has a long history of being something of a British preoccupation (Edwards, 1944). Some would suggest that it has done so to the point where it has suppressed much design innovation, particularly in housing (Moro, 1958; Manser, 1993).

Professional interest in design is partly focused on reusing brownfield land, increasing emphasis upon densities that will support public transport, and mixing uses so that facilities and employment are more accessible on foot.

More recently, professional interest is reflected in concern about the reuse of brownfield land, increasing emphasis upon densities that will support public transport, mixed use, and road and parking patterns that calm traffic, increase safety and security and improve connectivity. There has also been a return to questions of space in and around the home, and the qualities of that space through a consideration of factors such as light and microclimate, flexibility and adaptability and lifetime use (see DTLR/CABE, 2001: 62-9). The Government's commitment to the home zone concept also reflects a desire to encourage concern for the social aspects of residential layout and public space use (Biddulph, 2001; IHIE 2002). Meanwhile, the market continues to be left to set space standards within the home.

House builders argue, however, that they are best placed to understand the house-buying public's housing preferences as they regularly research it and consumers' decisions to purchase. They can suggest that professional judgment about housing forms and styles, street layouts and the public realm, neighbourhood designs and preferred locations can sometimes contradict these preferences. In particular, professionals might argue that policy in the public interest should encourage redevelopment in urban centres, at higher densities and with lower parking standards, provide greater levels of permeability and include development of more sustainable forms of housing. Conversely, house builders may argue that consumers do not relate directly to issues affecting the wider public 'good' and instead prefer low density schemes in suburban or greenfield locations with additional parking spaces and greater levels of exclusivity. Similarly, whilst consumers may generally agree with the need to move towards more sustainable lifestyles, they are less willing to pay for aspects of a house which promote energy conservation.

Such contradictions however remain contested, and the case can be made that although particular groups may not agree on what they like about schemes, such matters do not necessarily contradict. Instead, within and between groups, there is significant scope for overlap as preferences evolve. Emerging niche markets illustrate that there are opportunities – and successful examples – where schemes which would be seen to reflect professional tastes can also conform to some areas of public aspiration, and ultimately influence in a positive way the value that is accrued by developers from better designed schemes (CABE, 2003). Preferences may therefore be shaped by group affiliations, and people or groups might express either a concern to reflect issues of private or public interest. There is evidence however that preferences continue to evolve and that such preferences do not necessarily contradict.

Such matters are relevant to any consideration of awards for housing design for two reasons. Firstly, it is important to judge whose notions of good design are being promulgated, and whether there are any tensions emerging between groups as a result. Secondly, it is useful to consider the extent to which there is a distinction between, for example, professional, consumer and public tastes, and whether involvement in awards is affected by judgements about the types of interest that are being represented. Will house builders be put off entering awards that they feel are focussing on professional notions of good design, as opposed to what they regard as their consumers' concerns? In the next section research specifically into design competitions is considered. This seeks to uncover the bases of decision making, the quality criteria that are being used, whose values are being promulgated through the judging processes, as well as the public reactions to the schemes that are winning design awards.

Architectural competitions and the judgement of design quality

Vischer and Cooper Marcus (1986) analysed the biannual design award run by the Canadian Housing Design Council in 1981 to consider the extent to which designers', consumers' and judges' evaluative criteria overlapped. Their findings show a certain degree of overlap in criteria adopted by the groups, especially related to the overall look of the scheme. Residents were more concerned with issues about construction quality, security and privacy. Designers expressed concern about a variety of issues, although they would particularly comment on the design process, public involvement and why the schemes were good given the constraints of funding or the site. The judges seemed, in contrast, to be most willing to award architectural innovation: "*The jury's emphasis on innovation can be attributed partly to a pervasive value in the architectural profession – rewarding forms and ideas that are new and non-derivative...*" (Vischer and Cooper Marcus, 1986: 80).

Nasar (2000) reviews a range of international design commissions and re-runs them with contemporary lay and professional audiences to show radically different results. In exploring how "meanings matter" in competitions, and whose meanings are indeed influential, Nasar identifies seven key environmental features, which he argues are the essential means of delivering emotional meaning and which subsequently influence public preference. These are:

- Diversity (visual richness) – the number of noticeably different elements in a scene.
- Order – the degree to which a building looks organised and its parts unified.
- Openness – the openness and definition of the vista.

- Naturalness – the prominence of vegetation and water.
- Upkeep – the perceived level of maintenance and cleanliness of a place.
- Historical significance – the perception of historical significance of a place and traditional-style buildings.
- Liveable space – the presence of people.

Nasar argues that people place different emphases on each of these qualities in different localities, so that in residential areas relaxation may be more important than excitement or the presence of many people might be regarded as less desirable than in other types of environment. He argues that competition entries can be measured against these preferred features and that this would yield more predictable and more widely shared positive results. To achieve this Nasar suggests that guidelines for competitions and awards should be developed which either reflect these criteria or which would encourage interviewing the affected public in order to define the most sought-after qualities. Additionally, he suggests a "*pre-jury evaluation*" to provide insight into the public responses to different designs. Nasar strongly advocates defining appropriate criteria and managing the jury process carefully, both in terms of selecting appropriate professional adviser(s) and jury ("*pick the jury and you have decided the winners*"). Criteria used must ensure an appropriate balance of perspectives and expertise; there should also be discipline in the assessment process, including testing the schemes against a wider audience and keeping a formal record of proceedings.

How this previous thinking informs this study

Research on housing preference, and how preference is shaped amongst different actors, clearly demonstrates the need to be aware of the complexities involved in judgements about housing design quality. Who is making the judgement, to what end, and on what values are those judgements based? Tensions can be seen between professional and lay tastes, and between those who merely visit or pass by a scheme and those who buy into and live in it. Of course it is not simply a case of difference between groups; great diversity of opinion can be seen within groups. Architects can be deeply divided on the importance of context and tradition in housing design, about vernacular versus modernist responses to layout and styling. Planning, landscape and engineering professionals have different perspectives on, and responsibilities for, the final product.

House builders can argue that professional preoccupations more generally fail to appreciate the real interests of consumers, although there is some evidence to suggest that such an argument is not so easily made as consumer preferences evolve and many award winning schemes are happily lived in. Given this complexity, it can be argued that design awards should either acknowledge that they reflect and promote a particular agenda, or should be designed to enable judging – both in terms of criteria and process – to take into account differences of preference.

Factors influencing preference evolve, and an increasingly pluralist society exposed to an ever-wider design palette should perhaps lead to a wider choice of living environments and dwelling types. The nature of design awards should also reflect this. Nevertheless, it should also be acknowledged that certain tensions will continue to persist. Tensions between a concern to reflect professional versus consumer preference will continue to be apparent and particularly in the extent to which architects' interest in innovation is weighted relative to criteria regarded as more in tune with consumer interest and concerns. For example, a concern about the national imposition of 'a preferred style' on distinctive localities might persist if national judges tell local people what types of housing should be winning awards in their localities. Judging may also reflect tensions between traditional concerns about vernacular versus modern approaches to housing and residential design. Similarly, tensions may become apparent between a perceived consumer desire to live in low density suburban environments and professional commitment to raise densities and subsequently consider new forms of housing.

Chapter 3

Government Policy for Housing Design and Layout

Chapter 3

Government Policy for Housing Design and Layout

Government interest in issues of housing design and layout reflects the extent to which both issues are in the public interest. This section reviews national policy documents to consider which aspects of housing design and layout the Government may wish to influence through its involvement in sponsoring awards such as the *Housing Design Awards*. Clearly awards represent only one of the mechanisms via which the Government can try and influence the work of private developers, social housing providers and designers. Initially, therefore, the range of mechanisms for influencing design and layout are presented so that the position, role and significance of design awards can be appreciated.

Mechanisms for encouraging greater attention to housing design

The Government has a number of powers and mechanisms available to it, in particular through the planning process, to control and guide the design of housing schemes.

Specific mechanisms that exist to *control* development include:

- Approved development plan policies related to housing design (see Punter and Carmona, 1997; Carmona, Punter and Chapman, 2002).
- Local Authority policies for the adoption of particular forms of highway, typically based on Government advice in Design Bulletin 32, Residential Roads and Footpaths (DoE/DoT, 1992) and Places, Streets and Movement (DETR, 1998) (see below).
- Powers associated with the sale of land where design requirements might be more firmly enforced through a development brief.

Specific mechanisms that exist to *guide* development include:

- Planning Policy Guidance (and related documents) which refer to substantive residential design issues (see below).
- Local authority advice in the form of residential design guidance (see Essex Planning Officers Association, 1997; Bristol City Council 1998; Kent Association of Local Authorities, 2000).
- Design advice in site specific development briefs, where the local authority does not also own the land.
- Negotiations undertaken through the process of gaining planning permission.

In addition, the Government can also *encourage* improved design in housing development. Specific mechanisms that exist to encourage improved design include:

- Powers to grant aid/funding for schemes where designs conform to a particular standard (e.g. via English Partnerships or the Housing Corporation).
- Department for Culture, Media and Sport (DCMS) and Office of the Deputy Prime Minister (ODPM) sponsorship of the Commission for Architecture and the Built Environment (CABE), an organisation which aims to improve people's quality of life through the promotion of better design of buildings, spaces and places. Following recommendations made by the Urban Green Spaces Taskforce, a new unit within CABE – CABE Space – was launched in May 2003 to act as a champion for improving and promoting quality public and green space.
- By sponsoring events or publications related to housing design.
- By initiating or sponsoring design competitions relating to specific development opportunities.
- By sponsoring awards for housing design.

Within this context and when compared to the other mechanisms, awards for housing design are useful for three main reasons:

- they represent a reward for a completed initiative;
- they draw attention to good practice in relation to specific themes and issues highlighted by the award, and
- that good practice can then be more widely disseminated to an interested audience.

The policy context for encouraging attention to housing design

An urban renaissance

The wider context for supporting awards for housing design is provided by the Urban White Paper, *Our Towns and Cities: the future. Delivering an urban renaissance* (DTLR, 2000). This document links the design of housing to a range of interrelated government objectives. These include encouraging neighbourhood renewal, providing patterns of development that allow or encourage environmentally benign lifestyles, maintaining public health, providing adequate affordable housing, and ensuring adequate provision of suitable housing for people with special needs. The emphasis placed in the Urban White Paper upon the vital role of good urban design in general, and of good residential design in particular, clearly reflects the conclusions of the Urban Task Force report *Towards an Urban Renaissance* (1999).

Sustainable Communities

More recently, the Government has published *Sustainable Communities: building for the future* (ODPM, 2003) which presents a programme of action to tackle pressing problems and raise the quality of life in all communities. It builds upon actions already embarked upon, and marks a step change in policy delivery. The key elements of the action plan include:

- the promotion of significant development areas in Thames Gateway, Milton Keynes-South Midlands, Ashford and London-Stansted-Cambridge (LSC);
- creating the conditions in which private house builders will build more homes of the right type in the right places;
- encouraging a better mix of housing in new developments, including more homes that are affordable for those on modest incomes, and especially key workers;
- overcoming low demand and abandonment in the North and Midlands; and
- supporting the Commission for Architecture and the Built Environment (CABE) to drive up design standards

The provision of new homes is central to the plan, and it is clear that ensuring quality of design, and thereby regaining public trust in new homes is a key Government objective:

> There is a clear demand from the public for higher quality homes and neighbourhoods. In a recent survey, 85% of the population stated that they were interested in or very interested in the state of the local environment, and recognised the relationship with their overall quality of life. By contrast, a large majority did not consider that current housing developments were well designed.
>
> **ODPM, 2003**

Planning Policy Guidance

The specific policy context in which the present consideration of the role of housing design awards is to be considered is established by relevant Planning Policy Guidance (as set out in *PPG1: General Policy and Principles, PPG 3: Housing*, and *PPG15: Planning and the Historic Environment*. These Guidance Notes acknowledge the significance of good urban design in achieving mixed uses, sustainable forms of development, urban regeneration, protecting and enhancing listed buildings and their settings and enhancing the character of conservation areas respectively. A new Planning Policy Statement (PPS)1 – *Creating Sustainable Communities* was out for consultation at the time of going to print on this report. The PPS will replace PPG1.

PPG1 (DTLR, 2001) establishes that the appearance of a proposed development and its relationship to its surroundings constitute material planning considerations. In addition it also points out that design polices and guidance should avoid stifling responsible innovation, originality or initiative; things that awards for housing design should, in contrast, encourage. Annex A provides advice to Local Authorities on how design issues should be dealt with through the development planning process.

PPG3 (DETR, 2000) seeks to promote good design in new housing developments in order to create attractive, high-quality living environments in which people will choose to live. The sustainable development and mixed-use objectives of PPG 1 are extended to include the creation of mixed communities. This influences the type and size of housing, as well as issues such as affordability, meeting the needs of special groups, and avoiding housing development which makes inefficient use of land. Local planning authorities are urged to encourage good design to promote sustainable development. Emphasis is placed on quality and designing places for people, with good design and layout of new development helping to achieve Government objectives of making the best use of previously-developed land and improving the quality and attractiveness of residential areas.

PPG15 (DoE/DoNH, 1994) promotes the need for due consideration in the design of new development next to historic buildings, and that new buildings should be 'woven into the fabric' of the environment and community. Importantly, the guidance suggests that new buildings should not necessarily mimic their older neighbours. It also notes that the Royal Fine Arts Commission is to be consulted on developments affecting conservation areas of national importance.

National Good Practice Design Guidance

A feature of recent national Planning Policy Guidance has been the use of cross-referencing to further guidance and advice on design endorsed by the Government. Of particular relevance to housing design are:

- *Places, Streets and Movement: A Companion Guide to Design Bulletin 32 (DB32) Residential Roads and Footpaths* (DETR, 1998).

This guide encourages a greater emphasis upon place, community and context in the design of housing layouts, and is designed to ensure that DB32 is used more imaginatively than has previously been the case. Above all, the Companion Guide seeks to promote a balanced approach in which good urban design promotes sustainable development and facilitates an integrated transport strategy.

- *By Design: Urban design in the planning system: towards better practice* (DTLR, 2000).

This report establishes good urban design as fundamental to the objective of urban renaissance, and puts forward practical recommendations to promote higher standards in urban design. The guide encourages a move away from a negative reliance on standards towards a more positive emphasis on performance criteria. The guide also identifies seven mutually reinforcing objectives underlying good urban design, which are related to aspects of development form.

- *By Design: Better places to live: A companion guide to PPG3* (DTLR/CABE, 2001).

Applying the fundamental principles of good design outlined in *By Design* (DTLR, 2000) to residential environments, this guide encourages and challenges local authorities and developers to think more imaginatively about housing design and layout. It seeks to promote greater flair in creating better places in which to live. Its focus is the urban design principles and approaches that underpin successful housing. It is relevant to the whole spectrum of housing, but a major focus is housing within the density range of 30-50 dwellings per hectare. The guide seeks to promote a holistic view of the design of residential environments and identifies fourteen attributes of successful housing. These attributes are considered in the context of case studies which consider context, principles of movement, housing mix and neighbourhood and aspects of housing layout and urban form.

- *The Urban Design Compendium* (Llewelyn-Davies, 2000).

The Urban Design Compendium, sponsored by English Partnerships and the Housing Corporation was developed alongside DTLR's *By Design* publication, and addresses many of the urban design issues associated with the regeneration and development problems identified by the Urban Task Force. Guidance contained within the Compendium relates to both the 'product' and 'process' of urban design, noting that this is not a 'tick-box' exercise. The substance of the advice is strongly complementary to that in the other guides outlined above.

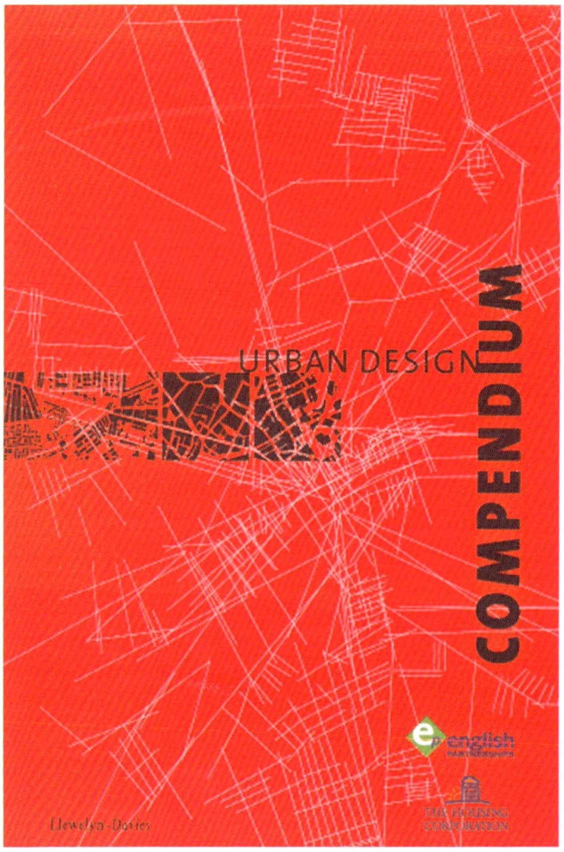

By Design: Better places to live and the Urban Design Compendium are two pieces of national good practice design guidance that embrace housing layout and design issues

The Commission for Architecture and the Built Environment (CABE)

CABE was established in 1999 to act as champion for architecture and design in the built environment in England. The organisation exists to promote and encourage high standards in the design of buildings and the spaces between them by helping and offering advice to all those who create, manage and use the built environment.

With a particular focus on housing design, CABE is working with the House Builders Federation and Civic Trust through the *Building for Life* initiative. The aims of the initiative are threefold. The first is to gather examples of exemplar housing design, both at home and abroad, and explain to the house building industry why these designs work so well and how they can be learnt from. The second is to better understand the aspirations of people buying homes so that the design of new housing is more attractive to them. Thirdly, the initiative will identify barriers to designing quality new homes and campaign to remove them.

This brief review of the tools and mechanisms available to promote and support better housing design illustrates how and why it is an important consideration for the Government if it is to ensure delivery of key areas of policy. In particular, it is central to its commitments to achieve an urban renaissance, to meeting demand for housing in the South East, overcoming low demand in parts of the North and Midlands, and creating balanced and sustainable communities with adequate provision of affordable housing.

Through the planning process Government has established a framework for guiding and controlling the form of housing, and has provided both advice and information to both planning authorities and developers as to the forms of housing that it seeks to endorse. Government support for awards for housing design makes a very specific contribution to these goals away from the planning arena. In particular, the ability for design awards to both reward and encourage is distinctive, whilst their ability to bring forward design exemplars for others to consider and learn from is also important.

Chapter 4

Awards for Housing Design

Chapter 4
Awards for Housing Design

Awards relating to the design and layout of housing embrace a wide variety of factors relating to perceived housing quality. As discussed in Chapter 2, particular social groups may demonstrate differing housing design preferences, and such differences carry through in terms of the complexities involved in judging and assessing what may constitute good design. Therefore as may be expected, a large number of awards exist across the UK concerned with evaluations of different aspects of housing development. Whilst many of these (often featuring in the property sections of national or regional newspapers) include aspects of housing design, other attributes of housing (such as construction technology or building management) play an equal or perhaps more dominant role in the award evaluation. Moreover, some awards may include housing development, but may not have housing as the central focus of concern.

As part of the initial review, all those awards in which housing design plays a dominant role were identified. The list below represents the range of awards available to architects and housing developers. The range of awards presented reflects the wide range of interests and concerns to all those involved in the housing design and development process, including awards that reward quality in construction, sales and marketing or after sales services. Although not design related, many of these awards can be highly valued by housing developers.

- British Housebuilder Awards
- Building Homes Quality Award
- BURA Award for Best Practice in Regeneration
- BURA Charitable Trust Community Regeneration Awards
- Civic Trust Awards
- Europan (A design competition)
- Evening Standard New Homes Awards
- Greenleaf Awards
- Housing Design Awards
- Inside Housing UK Housing Awards
- National Home Builder Design Awards
- National Home Improvement Council Award
- RTPI Planning Awards
- Saltire Society Housing Design Awards
- Welsh Housing Design Awards
- What House? Awards

Whilst each can be seen to serve a particular niche or focus on a particular interest, it clearly would have been too unwieldy to conduct a review across all these award programmes. The key focus of this research has therefore been to identify and consider in more detail the principal awards that promote and encourage better housing design. Upon inspection of the competition rules, categories and judging criteria, the initial list was shortened to exclude those awards which did not have housing as a central focus, which did not apply across the spectrum of housing providers, or in which housing design was not a dominant concern. Awards that apply to a specific regional context within England (largely defined by local newspaper catchment areas) were excluded. An exception was made with the *Evening*

Standard New Homes Awards which covers the South East of England, since housing provision in this region constitutes such a significant proportion of the English total. Awards solely operating outside England (for example the *Saltire Design Awards*) were also excluded. This sieving process led to a focus for this piece of research on the following six awards:

- National Home Builder Design Awards
- Civic Trust Awards
- Evening Standard New Homes Awards
- Greenleaf Awards
- Housing Design Awards
- What House? Awards

It is important to note that knowledge of both the general housing awards as well as the more specific awards for housing design was not equally distributed amongst the case study interviewees, with even major house builders seeming to be unclear about the whole range of awards available to them. The possible reasons for this shall be discussed in more detail in Chapter 5.

Selected design awards and their focus

National Home Builder Design Awards

This is one of the two (alongside the *What House? Awards*) long-established awards for housing design sponsored by a national newspaper. Established in 1981, the awards were recently restructured in 1999 to apply exclusively to schemes designed by architects working for private house builders producing housing for sale. This renewed focus arose from a concern that the competition required clearer criteria and a more selective evaluation process. The awards feature a large number of detailed categories encompassing a wide range of housing design and clearly focus upon the house building industry itself. Newer categories in the awards include restoration and conversion, brownfield residential development and partnership development.

Interesting features of the awards

The awards:

- are likely to be attractive to the house building industry (including partnership developers);
- have involved restructuring to incorporate aspects of the professional/lay evaluation concerns noted in Chapter Two;
- include categories that encompass both housing design and urban regeneration.

Civic Trust Awards

Established in 1959, the unique feature of the *Civic Trust Awards* is that they take into account the benefit each project brings to its local area as well as considering the quality of its design. Unlike other awards, entries are organised with the help of participating local authorities, with assessors nominated by the Civic Trust. Each assessor team is lead by an architect, and also includes a local planner, a lay adviser and, where possible, an expert in disabled access. Awards are given on a two-year cycle, alternating between major city areas and countryside contexts. A range of awards is given, including for urban and rural design, sustainability, landscape, partnership and for town/city centres. Awards are for outstanding exemplars, but also include commendations and mentions, the latter involving a clear and significant benefit for the local community.

Interesting features of the awards

The awards:

- are well-established, embracing a wide range of features of urban and housing design;
- involve the mediation of the public sector in their administration (as a 'balance' to those awards oriented to the private sector);
- relate urban design to issues of social inclusion and community benefit (i.e. deal with aspects of both housing design and urban renaissance).

Evening Standard New Homes Awards

Focused upon London and the South East and as such a regional rather than national award, the Evening Standard New Homes Award is oriented towards the largest housing market in England. The Awards are essentially geared towards the private house building industry, but categories have been expanded to include housing association development. Two sets of awards are given for each category, distinguishing between house builders constructing over 100 units a year and those building under 100 units. Judges include lay persons selected from readers of the sponsoring newspaper.

Interesting features of the awards

The awards:

- are likely to involve a majority of the house building firms active in the most dynamic housing market in England;
- involve lay and expert judges;
- include a wide variety of categories relating to housing design, which have been expanded to include aspects of sustainable development.

Greenleaf Awards

Established in 1986 by the New Homes Marketing Board (House Builders' Federation), the unique focus of these awards is the space between buildings, recognising the best in new housing landscapes. Judges are not professionally led, the landscape being judged for 'liveability'. Awards are given to schemes that conform to a standard and schemes must be at least three years' old so that the landscape has had a chance to become established.

Interesting features of the Awards

The awards:

- relate to an area of housing design often not 'foregrounded' in other awards;
- involve assessment which is not professionally-led;
- involve rewarding schemes which meet specified standards.

Housing Design Awards

The original Government sponsored awards scheme, relating only to the design and layout of public housing, ran from 1950 to 1955. In 1960, the awards were reconstituted as the Ministry of Housing and Local Government (later Department of Environment) Good Design in Housing Awards to cover both public and private sector housing and became sponsored jointly by the Royal Institute of British Architects (RIBA). In 1981, the National House Building Council (NHBC) became a sponsor, the awards moved to a biennial timetable and the separate public/private sector categories were abolished. Since the mid-1990s schemes at project stage as well as completed schemes, have been included. Sponsors of the current *Housing Design Awards* include the Office of the Deputy Prime Minister (ODPM), the NHBC, RIBA and the Royal Town Planning Institute (RTPI).

The awards are open to developments of four or more dwellings, and may include a non-residential component as long as housing constitutes the major element. Awards are made to completed schemes and to projects that have detailed planning permission, but which are not yet complete. Special awards were introduced in 2000 – including a Regeneration Award, Greenfield Award and Sustainability Award.

Interesting features of the awards

The awards:

- are the longest-established, with an 'establishment' pedigree;
- have rigorously-specified criteria, relating to excellence in housing design;
- are associated with (amongst others) the RIBA and RTPI, and hence would shed further light on the professional/lay dimension of housing design evaluation noted in Chapter 2.

What House? Awards

The Daily Telegraph

Established in 1981, the What House? Awards are very much focused towards the private house building industry. There are a large number of categories of awards, which include provision for large (1000 units +), medium (100-999 units) and small (less than 100 units) firms.

Interesting features of the awards

The awards:
- offer an alternative outlet to the private house building industry to the National Home Builder Design Awards;
- have expanded categories to embrace aspects of the urban renaissance agenda;
- involve architects in judging schemes that represent different market contexts.

Categories have been expanded to reflect brownfield development, partnership development and sustainability (principally concerned with energy-saving measures). The judging panel comprises architects and property journalists. Awards include 'gold', 'silver' and 'bronze' in each category.

More comprehensive details of each of these awards are presented as Appendix 1. Summary details of the other housing-related awards can be found as Appendix 2. There are of course a number of similarities between the six key awards briefly introduced above. There are, however, particular foci and emphases that can be drawn out.

- Three of these awards, the *National Home Builder Design Awards*, the *Evening Standard New Homes Awards* and the *What House? Awards* largely reflect the concerns of the house building industry with housing design issues and seek to combine an interest in improved design with market and consumer concerns.
- One award (the *Greenleaf Award*) focuses upon landscape issues from the perspective of 'liveability'. Two awards (*Housing Design Award* and *Civic Trust Awards*) have a public-sector/professional orientation towards promoting urban design and housing design excellence.
- Two of the awards (*National Home Builder Design Awards* and *What House? Awards*) are clearly competitor awards, serving a national market for house builders in a highly competitive environment. The *Evening Standard New Homes Awards* serve a similar function for London and the Southeast.

Notably, all have expanded their categories in recent years to include aspects of policy concern including urban sustainability, renaissance and partnership development, and each has widened its focus to embrace aspects of design beyond the dwelling itself. To this extent, some of the categories in each award are now more competitive with other awards (landscape categories, for instance, competing with the *Greenleaf Awards*, though often using different criteria).

Analysing the awards for housing design

Using awards documentation and interviews with the organisers of the selected awards, it was possible to compare and contrast the different approaches that are taken to award giving. In particular it was possible to consider:

- judging procedures;
- judging criteria;
- judging categories;
- approaches to dissemination of award winners;
- which types of house builder tend to be involved in which awards, and also whether certain development schemes appear in more than one award programme.

Information was requested for awards given between 1997 and 2002. However, there were a number of factors that hindered collation of material:

- Not all competition organisers maintain full records relating to their awards, and for a number of the competitions considered, full information was not available.
- Information on award winners is also often more comprehensive than that relating to competition participants.
- Many competitions change their programmes significantly on an annual basis, making clear comparisons over time problematic.

Nevertheless, the following analysis offers the first systemic review of the major housing design award competitions in England between 1997 and 2002.

Judging procedures

Whilst approaches to judging vary, most approaches appear to be rigorous: only one award, for example, does not include site visits. Judging does, however, vary as to the extent to which actual consumers of housing are engaged in the judging process.

Civic Trust Awards

The judging procedure takes 5 months. Entries are organised with the help of participating local authorities (LAs). A member of the LA arranges for a judging team to visit the entries. A team of assessors for each area is nominated by the Civic Trust, and lead by an architect with no links to the area. The team also includes a local planner, a lay adviser and an expert in disabled access. Views of local residents on the scheme are sought and the local press is reviewed. The team leader writes a report. A national panel of expert judges reviews the recommendations. The group meets to ensure comparability of judging from across the UK. The group also agrees which special awards will be given unless it is felt further specialist judging for those awards is required.

Evening Standard New Homes Awards

The team of judges selected by the newspaper (approximately 8) assesses all the entries and establishes a short list in each category. Short listed properties are visited and 10 schemes in each category are selected. Evening Standard readers are invited to form part of a judging panel responsible for different categories. The readers then choose a 'winner' and a 'highly commended' for each category.

Greenleaf Awards

Schemes are initially judged on paper. Schemes not established for 3 years and which do not show care and consideration for the landscape are set aside. For proceeding to the next stage of judging, a more detailed application form is sent out to those schemes selected. A group of lay persons, representing a range of sponsors, are trained in the judging criteria. Members of this group visit every site and complete a structured report. A group of main judges then see these submitted comments and compare these with the photos to agree recommendations. For the specialist awards, someone with more specialist knowledge will then visit the sites.

Housing Design Awards

Each sponsor contributes two assessors to the judging panel. Entries are sifted to pull out clear 'non-starters.' Using a set of criteria, the panel then selects and discusses schemes in determining which should be selected for the exhibition (typically about 70 boards). From the projects chosen for exhibition, 8-10 project award schemes are selected and 14-16 completed schemes are chosen for visiting. Assessors also then select candidates for the special awards. Half of the panel visit the short-listed completed schemes over two 2-day visits to determine the final 3-6 winners. Once awards are fixed, special awards are also determined – a task normally undertaken by the special awards sponsors.

National Home Builder Design Awards

The editor of Planahome magazine appoints 6-8 judges. A balance is sought between architects and 'informed' lay people. Typically, the 'lay' jurors express their preferences first. The judging panel carries out a preliminary shortlisting, deleting those not up to scratch. A collective view is reached on a second round, and successful schemes are then rated to provide (as a third round) a short list, on the basis of which the final decisions in relation to each category are decided. Most schemes are not visited, except those schemes shortlisted for the Manser Medal.

What House? Awards

The judges produce a shortlist of the best entries in each category. Shortlisted schemes in each category are then visited and a report written before the judges make their final decision. Where judges cannot make a clear cut decision entrants may share a prize.

Further details of the judging procedures are presented in Appendix 1.

Two broad types of judging procedure can be identified. Awards such as the *Housing Design Awards, National Home Builder Design Awards and What House? Awards* tend to focus on informed lay and professional views of the housing product, whereas the *Evening Standard Awards* and *Civic Trust Awards* formally seek and use unaffected lay opinion. The *Greenleaf Awards* span the two approaches, using lay judges but adopting specific criteria provided by the award organisers.

Chapter Two highlighted some of the tensions between consumer, public and professional evaluation of housing design. As discussed, there is an important distinction to be made between general public evaluations and house buyer preferences, and it is not always obvious from the criteria deployed in competitions that this distinction is clearly made, nor what the consequences are in terms of scheme selection for awards.

Judging criteria

It has been noted previously that many of the housing design awards have sought to engage and reflect design issues emerging through Government guidance, the creation of CABE and the Urban White Paper. In order to provide a brief view as to how judging criteria respond to themes associated with the urban renaissance, Figure 1 considers judging criteria for the six key housing design awards considered in detail against areas or themes derived from the DTLR/CABE (2001) best practice guide '*By design: Better Places to Live*'. Most awards have some criteria approximating these, although the extent to which they are used seems to vary in practice, with a number of interviewees referring to matters such a 'liveability' or 'gut feelings' as being the key to recommending schemes in practice. Most awards have only selective criteria, or they focus on a specific aspect under a general heading. For example, references to 'space in and around the home' in the Civic Trust Awards tend to focus on provisions for disabled access.

Interestingly, most criteria referred to in supporting awards documentation relate to themes to judge, rather than standards to reach. Typically, criteria emphasise nouns such as 'landscaping' rather than adjectives such as 'beautiful' 'maintained' or 'innovative' describing the quality of that noun. It is also noteworthy that 'housing mix and neighbourhood' does not (yet) feature as an explicit criterion in any award, despite featuring as an important aspect of recent housing design policy.

The *Greenleaf Awards, Housing Design Awards* and *Civic Trust Awards* seem to have the most explicit criteria, and these awards also provide the more extensive advice to judges. Interestingly, and as stated previously, the volume house builders interviewed for this research felt that the judging criteria adopted in the *What House? Awards* were the most consumer-oriented. By contrast, the volume house builders felt that the *Housing Design Awards* only recognise schemes that are of a modern style. Neither of these views appear to be supported by the judging criteria, although it may be the case that such preferences become apparent in the application of those criteria.

Figure 1: Comparison of judging criteria adopted in awards for housing design

	National Home Builders Design Awards: Housing Dev. of the year category	Civic Trust Awards	Evening Standard New Homes Awards The Best New Development category	Greenleaf Awards	Housing Design Awards	*What House? Awards* Best Development category
General Statements	"Outstanding examples of contemporary design, layout and landscaping taking into account local conditions, materials and other factors"	None	"The best development achieving the highest standard in layout, landscaping, quality of construction and overall use of the site…"	None	None	"Judges will be looking for a stylish mix of original designs, top quality site layout and landscaping, and variety of external elevations and plenty of design flair inside and out."
Relationship with the Scheme Context	None	The way a scheme relates to its surroundings	None	Does the development fit in with the style and scale of the surroundings, and does the treatment strengthen physical links and reinforce local character?	Relation to surroundings and neighbourhood Response to site constraints and opportunities.	None
Movement Framework	None	None	None	None	Planning of roads and footpaths.	None
Housing Mix and Neighbourhood	None	None	None	None	None	None

Housing Layout and Urban Form	Layout	None	None	Provision of communal space: seating in the sun, accessible water features, shared play areas? Planting to help with enclosure, shelter, screening and the provision of a green setting. Views in and out of the site and along main route paths. Privacy and enclosure	Garages and car parking Safety, security and accessibility	None
Space in and around the home	Overall appearance External and interior design. Design of the dwellings. Landscaping.	The provision made for disabled access	None	None	External appearance	None
Maintenance and finish	None	Maintenance and weathering	None	Designed for ease of maintenance. High quality hard landscape	Internal planning Finishes, detailing, workmanship	None
Other criteria	"Every entry must have been designed by a named architect on the current register of the Architect's Registration Board, the RIBA or an equiv-alent overseas body"	Local opinion History of the site and that historic buildings/ landscapes have not been destroyed	None	Sustained aftercare. Care in construction. Conservation of existing features.	Sustainability in construction	None

Based upon current best practice guidance, **Figure 1** indicates that the *Housing Design Awards*, the *Civic Trust Awards* and *Greenleaf Awards* utilisation of explicit criteria can be more readily matched with the major components of urban and housing design accepted amongst professional urban designers. In the case of the other awards this is more difficult to establish, since judgements seem to be made 'in the round' or upon different (not always explicit) criteria. This may reflect the differences between professional and house builder orientations to housing design noted in Chapter 2.

Types of award and judging categories

The categories adopted in the six selected awards are shown in **Figure 2**. They appear to have changed quite frequently in recent years, although general trends in category types can be identified. Typically there are four types of award given within the programmes:

- **Exemplar Awards**: Awards for exceptional schemes that highlight generally exemplary and innovative design. These awards are particularly attractive to designers who seek professional kudos and recognition from innovative designs, but they are also interesting to developers in terms of transferable lessons when useful exemplars win. Most developers may be put off entering such awards, however, as much housing development is not particularly innovative.

- **Benchmark Awards**: These are awards that reward designs that meet a particular standard and tend to be more attractive to developers as success is more certain and hence the effort is considered to be worthwhile. Others suggest, however, that such awards are devalued as they are sometimes too easy to win.

- **Category Awards**: Rewarding exceptional schemes in a particular category, these awards are valued by both developers and designers, especially where categories are aligned to particular products (for example starter homes, apartments or luxury houses). Categories can also reflect particular issues such as reusing brownfield land, development at higher densities and sustainability. Sometimes more than one award is given in each category (for example Gold, Silver or Bronze awards are given in each category of the *What House? Awards*). Some professionals would question the value of such awards, however, if there are too many categories and, like benchmark schemes, they become too easy to win. Organisations running the awards indicated that most categories seem to be well supported with a good selection of entries, and in particular schemes that conform to the urban renaissance agenda such as schemes on brownfield land or at a high density (apartments). Less entries tend to be seen for greenfield sites and low energy/'eco' categories. In years where there are few entries organisers indicated they may not give awards in such categories – awards will not be given for the sake of giving awards.

- **Outright Winners**: A single scheme that is regarded as the outright annual winner of the programme. Being an outright winner is highly prized and particularly valued by the media, although some organisers questioned whether it was appropriate to have one outstanding scheme each year, given that forms of housing are quite diverse and both contexts and markets are very varied.

Typically award programmes contain a combination of the above types of award, and how they are combined is one of the principal ways in which the programmes become attractive or not to participants.

Figure 2: Judging Categories

	National Home Builders Design Awards 2002	Civic Trust Awards 2002	Evening Standard New Homes Awards 2002	Greenleaf Awards 2001	Housing Design Awards 2002	What House? Awards 2002
Number of housing design awards to a benchmark standard (Awards and Commendations)	0	17 (involving housing)	0	78	0	0
Number of special category awards and exemplars (Awards and Commendations)	38	0	33	6	19	72
Special Categories						
Project Award	•		• (4 categories)		•	•
Individual Dwellings	•		•			• (4 categories)
Scheme award/4 or more dwellings	(2 categories)		•		•	•
Apartment	•		•			•
Restoration and Conversion	•					•
New-Build on Brownfield Land	•			•		•
New-Build on Greenfield Land						
Sustainability	•	•				•
Landscape Design	•	•		•		•
Interior Design and Planning						•
Partnership Award	•	•	•			•
Other	Retirement	– Centre vision award – Access Award – Urban Design Award – Exterior Lighting Award – Rural Design Award – Welsh Regeneration Award	London Lifestyle Award	– National Urban Forestry Unit Award – Civic Trust Award – Wildlife and Wetlands Trust award – Tree Council Award	– Building for Life – House Builder – Housing Corporation	– House Builder of the Year – Best Volume – Best Medium – Best Small – Retirement – Exterior Design

Dissemination

Approaches to dissemination influence how well known the awards are and also therefore the willingness of the participants to become involved. Summary details of how awards are disseminated are presented in **Figure 3**. Awards are publicised through:

- a 'physical award' – a trophy, plaque or flags which can be located in offices and on site;
- exhibitions;
- profile in both regional and national newspapers;
- special publications and supplements;
- web sites.

Audience is clearly the determining factor in approaches taken with each award to dissemination. A key distinction within the awards focused upon in this research distinguishes between those underpinned by and aligned to a particular newspaper and those that are independent of a media 'parent'. The *National Home Builder Design Awards* and the *What House? Awards* are sponsored by national newspapers, the *Evening Standard New Homes Awards* are sponsored by a large regional newspaper and the Greenleaf Awards are sponsored by a newspaper proprietor who promotes the awards at a regional level. These awards rely on their dissemination being attractive to a lay public, which in itself makes the awards attractive to private house builders who regard that public as their key market. By contrast the *Civic Trust Awards* and *Housing Design Awards* will gain national newspaper profile, but the material that is used tends to have a professional focus and the awards are regarded as more elite.

Figure 3: Dissemination techniques

	National Home Builders Design Awards	Civic Trust Awards	Evening Standard New Homes Awards	Greenleaf Awards	Housing Design Awards	What House? Awards
Banner/certificate /plaque		•	•	•	•	•
Exhibition	•				•	
National newspaper profile						•
Local/Regional newspaper profile	•	•	•	•		
Special publications/ supplements	•	•			•	•
Specific web sites					•	

A number of 'spin off' publications are also generated by the various awards. The *National Home Builder Design Awards* has produced a full colour, glossy *Judges' Report* supported by commercial sponsorship through advertising, as well as an annual and glossy *Planahome: The book of house plans* magazine which is available in the high street. The publication is an impressive transformation of the Judges' Report into a more popular magazine, relying on additional revenue from advertising and selling in 2003 for £4.95.

The *Housing Design Awards* have previously been reported in a special supplement in the trade journal "Building". An annual award report is produced, including details of all of the schemes that have won awards or 'Worth a Detour' commendations. A publication has also been produced which presents details of some of the schemes presented in 2 years of the *Housing Design Awards* exhibition, *Pick of the Crop*. The *Housing Design Awards* also have a useful web site attached to the Design for Homes website which includes the same information as presented in the award reports.

The *Civic Trust Awards* are currently disseminated through a special supplement in Building magazine. There is also a website, but it only lists the award winners. The *Greenleaf Awards* also produce an annual award winners' brochure. *What House?* magazine publish an annual supplement and the awards are also covered in *Show House*, a trade magazine for the sales and marketing people in house building companies.

Effectiveness of dissemination is a key factor affecting the extent to which interests, and particularly house builders, will become aware of and subsequently involved in particular award programmes. The extent to which particular interests find these methods of dissemination both useful and interesting is discussed further in recommendations and conclusions arising from the research.

Which type of developers have won which types of award?

Using the information collected from the various awards referred to above **Figure 4** provides information about which types of company have won which awards. To do this the industry was divided into:

- volume house builders (>1000 units per year);
- medium sized house builders (500-1000 units per year);
- top small companies (46-500 units per year);
- other award winning companies (<46 units per year);
- social housing providers and other housing providers (schools, hospitals etc).

In Figure 4 the total number of times such a category of company appears in the winners for the different awards is indicated in bold, whilst the percentage in brackets indicates the proportion of award winners for that category within the particular award.

Figure 4: Categories of house builders/developers winning awards

	National Home Builders Design Award (1999-2002)	Civic Trust Award (1997-2002)	Evening Standard New Homes Awards (1998-2002)	Housing Design Awards (1997-2002)	Greenleaf Awards (1998-2001)
Large Companies +1000 completions (83,658 in 2001) 21 companies	36 (45.5%)	2 (4%)	12 (31.5%)	18 (15.5%)	143 (73%)
Medium Companies 500–1000 completions (9,006 in 2001) 12 companies	4 (5%)	0	0	0	11 (5.5%)
Top Small Companies 46 – 500 completions (10,334 in 2001) 46 companies	10 (12.5%)	1 (2%)	3 (8%)	3 (2.5%)	15 (7.5%)
Other "Award Winning Companies" (<46 completions) 89 companies	22 (28%)	9 (18.5%)	20 (52.5)	39 (34%)	27 (14%)
Social Housing Providers 36 organisations	2 (3%)	32 (65.5%)	3 (8%)	39 (34%)	0
Others eg Local Authorities/Universities 19 organisations	5 (6%)	5 (10%)	0	16 (14%)	0

The segmentation of house builders by size (in terms of number of units completed per annum) was undertaken for two reasons. Firstly, previous research (Nicol & Hooper, 1999; Hooper & Nicol, 1999; Hooper & Nicol, 2000; Adams & Watkins, 2002) indicates that such a distinction is closely correlated with thresholds reflecting the tendency of house building firms to use standardised designs and layouts. Secondly, some of the design competitions utilised categories specifically oriented to housing output. It seemed relevant, therefore, to explore whether this dimension had any influence upon housing design considerations across the competitions.

However, undertaking this analysis has been problematical for a number of reasons:

- From a systematic perspective, not all the awards cover the same time period, and some records are incomplete for certain years. In Figure 4, the years for which information is available is indicated under the award title.
- Some house building firms/housing providers have been the subject of mergers, acquisitions and take-overs or restructuring during the period under investigation.
- The terms used to describe award-winning schemes may fail to highlight the fact that an award-winning scheme is a building or phase of a larger award winning scheme, which might then infer some element of double counting.
- The categories or criteria may change over time.
- Importantly, the survey data only refers to award winners or commendations in all cases, and does not reflect on the larger range of companies that might become involved, as records are not kept or could not be accessed.

In comparing awards for housing design, therefore, it is important to note that quantitative indicators of success serve only to indicate broad patterns. As such, interpretation of Figure 4 should make allowance for the different numbers and types of awards across the competitions. In spite of these problems, however, this research represents the first systematic attempt to bring this type of information together, and the information that has been collected provides an important context for the overall discussion and analysis.

Figure 4 highlights a number of trends worthy of further consideration:

- Large companies account for most of the houses completed using 2001 figures (83, 658). Both medium sized and the top small companies roughly share the number of completed houses in the same year (9,006 and 10,334 respectively).
- Large companies are most involved with the *National Home Builder Design Awards* and the *Greenleaf Awards*, but they also figure quite highly in the *Evening Standard New Homes Awards*. Despite only having one year of results for the *What House? Awards* they also seem to focus on volume house builders (a finding confirmed by the interviews). This would indicate that the larger companies are most interested in (and indeed successful in) those competitions oriented towards the concerns of the house building industry. Most large companies feature in one or more of the award-winning categories in these competitions – some are regular winners in individual (or occasionally, several) competitions, whereas others appear on a more intermittent basis.
- Only a small number of medium sized and the top small house building companies appear in any of the awards programmes, but a large number (89) of very small companies figure comparatively well in all awards.
- A number of social housing providers (36) figure strongly in both the *Civic Trust Awards* and *Housing Design Awards*, but to date not at all in the *Greenleaf Awards*, and very little in the *National Home Builder Design Awards*.

House builders that frequently return to particular awards

550 awards given to 217 companies and organisations were reviewed in undertaking this research. Of these, only a small number have received awards from more than one award programme. There appear to be two main reasons for this. Firstly there is variation in the award categories which promotes diversity, including awards for individual houses as well as awards for schemes (e.g. *National Home Builder Design Awards* has 19 categories). Secondly there is variation in the focus for the different awards and variation in the extent to which certain types of organisation might participate.

To illustrate the absence of a large degree of cross-over between awards, since 1999 the *Housing Design Awards* and the *National Home Builder Design Awards* have shared only a number of private house builder schemes in common, as follows:

- The Point, Bristol (Crosby Homes)
- Deansgate Quay, Manchester (Crosby Homes)
- Century Court, Cheltenham (Crosby Homes)
- The Chronos Building, Mile End Road, London (Copthorn Homes)
- The Woodyard, Dulwich, London (Wates)
- Stonegate Place, Wye (Environ County Homes)
- Smithfield Buildings, Manchester (Urban Splash)

Since 1997 the *Housing Design Awards* and the *Civic Trust Awards* have shared only one private housebuilder scheme, but quite a significant number of Local Authority and social housing schemes, in common:

- Angel Town (Phase 1), Brixton, London (London Borough of Lambeth)
- CASPAR, Leeds (Joseph Rowntree Foundation)
- Focus Foyer, Birmingham (Focus Housing Group)
- Lotus Way, Clacton, Essex (Guinness Trust)
- Dalston Lane, Hackney, London (Peabody)
- Newington Green, London (Peabody)
- Murray Grove, London (Peabody)
- Gwynne Road, London (Ujima Housing Association)

The Chronos Building, Mile End Road London has won the Housing Design Awards and the National Home Builder Design Awards

Over the time period considered by the research, the *National Home Builder Awards* and the *Civic Trust Awards* share no award winning schemes in common.

Where 'overlapping' schemes such as the Point in Bristol are found which have been successful in both the *National Home Builder Design Awards* and the *Housing Design Awards*, it is interesting to compare supporting commentary from the judges:

The Point, Bristol

The Point in Bristol is described by the *Housing Design Award* judges as having:

> *"…a scale, sweep and assurance to this scheme which is all too rarely seen in private developments, and an attention to the quality of design extending from the internal planning, to the treatment of the public realm, which makes it an excellent augury for the future…a completely convincing piece of high density urban design…"*
> *(Anon, 2002)*

By contrast, the *National Home Builder Design Awards* Judges Report comments on the scheme:

> *"…This is an ingeniously conceived house design, giving urbanity and privacy and an amazing amount of private outdoor space, with not a hint of pastiche. It is a truly contemporary house that deserves to be adapted for use on other tight urban sites…"*
> *(Anon, 2002a)*

Figure 5 shows a 'sieve' of each of the 6 awards, illustrating those entrants awarded more than one award during the period for which information is available. In the case of the *Housing Design Awards*, detailed records since 1997 have been used; for other awards the period is shorter. It should be noted that the figures may reflect more than one award in any one year, and/or winners of awards plus 'commendations'. The recent spate of acquisitions and mergers within the house building industry inevitable complicates the picture. In certain cases (e.g. Taylor Woodrow), different components of a company in 2003 appear in three different awards (*Housing Design Award; Greenleaf; National Home Builder Design Awards*), but this may partially reflect the acquisition of another company which had won awards in earlier years.

There are a number of interesting features to note from **Figure 5**:

- Success in certain awards seems to be associated with particular kinds of housing provider. Thus the leading award winner in both the *Housing Design Awards* and the *Civic Trust Awards* is a social housing provider. In the case of the *Civic Trust Awards*, leading social housing providers may find the local authority involvement conducive to participation (and perhaps to partnering).
- Social housing providers feature significantly throughout the awards (including through partnership developments), and some awards have introduced specific categories for such schemes (e.g. *National Home Builder Design Awards* – Best Partnership Housing; *Evening Standard New Homes Awards* – Best New Development by a Housing Association; *What House? Awards* – Best Partnership Development).
- It is probable that the high level of return to awards in the case of the *Greenleaf Awards* reflects the fact that awards are given on the basis of conforming to a certain standard and are perhaps easier to attain.
- Companies building fewer than 50 dwellings in 2001 also feature significantly across all awards. Examples include Ballymore Properties, Octagon Developments and Wates.
- Companies specialising in urban regeneration also have a strong presence in the awards and have been repeatedly successful in a number of awards (e.g. Berkeley and its associate Crosby, Barratt and Countryside/Copthorn).

- Where companies have a regional organisation, a particular region is often over-represented in the awards (e.g. Barratt and Berkeley/Crosby).
- Many award-winning schemes involve joint ventures, consortia or partnerships, often involving a mix of private and public agencies.

Figure 5: House builders that frequently return to particular awards

National Home Builders Design Award (1999-2002)	Civic Trust Award (1997-2002)	Evening Standard New Homes Awards (2002)	Greenleaf Awards (1998-2001)	Housing Design Awards (1997-2002)	What House? Awards (2002)
Barratt (Southern) (8)	Peabody Trust (4)	Berkeley Homes (2)	Barratt (41)	Peabody Trust (10)	Countryside (4)
Crosby Homes (5)	Guinness Trust (3)	David Wilson Homes (2)	Redrow (35)	Crosby Homes (6)	Barratt (3)
Gleeson (5)	Ujima H.A. (2)	Millwood Designed Homes (2)	David Wilson Homes (16)	Urban Splash (6)	Bellway (3)
Berkeley (4)		Millgate Homes (2)	Beazer (12)	Guinness Trust (5)	Crest Nicholson (3)
Copthorn Homes (4)			McCarthy & Stone (7)	Ujima H A (4)	Laing (3)
Taylor Woodrow (4)			Bovis (6)	Joseph Rowntree Foundation (3)	Wilcon (3)
Emlor Homes (3)			Countryside (5)	Community H. A. (3)	Lindon (2)
Octagon Developments (3)			English Courtyard Association (4)	Copthorn Homes (2)	Nicholson Estates (2)
Ballymore Properties (2)			Fairclough (4)	Clare Hall (2)	Rosemullion (2)
Beechcroft Developments (2)			Jones Homes (4)	Coin Street Community Builders (2)	
Bewley Homes (2)			Stewart Milne Homes (3)	Countryside Properties (2)	
Environ Country Homes (2)			Wimpey (3)	Greenholme Investments (2)	
			Bryant (2)	Maritime HA (2)	
			Bewley (2)	Taywood (2)	
			Crest Nicholson (2)	Weltonvale Developments Ltd (2)	
			Grange Builders (2)		
			Alfred McAlpine (2)		
			Knapp New Homes (2)		
			Laing (2)		
			Morris (2)		
			Twigdon (2)		
			Wilcon (2)		

These findings suggest that there may be specific characteristics relating to the type and nature of awards which predispose house builders both to enter and be successful in them. This may relate to a wide range of possible factors, including (for example) orientation of firms to urban regeneration; degree of competition for complex sites; structure and management of the organisation; or utilisation of architectural expertise.

Key points from the awards analysis

From analysis of the key awards for housing design in England, a number of key points can be drawn out. These have guided the selection of schemes to provide case studies for interviews with builders and developers to be discussed in the next chapter. The key themes emerging from the analysis also help to illustrate how the awards vary in both form and focus.

Awards are given in 4 ways: (1) for a small number of exemplar schemes; (2) for a larger number of schemes that meet a certain standard in terms of quality; (3) for a range of exemplar schemes in specific categories; (4) for outright winners. Most awards programmes combine more than one approach. Architects seem to be attracted to (1), (3) and (4). Volume developers seem to be attracted to (2) and (4). Interest in the awards seems to reflect in many ways a series of vested interests. Architects are primarily interested in receiving either a *Civic Trust Award* or the *Housing Design Awards* 'Completed Scheme' award given their status with peers. To be attractive to private house builders it is felt that awards should be both worthwhile entering, i.e. probably leading to an award and therefore marketing potential, or providing exemplars that are useful for future practice.

Similarly, judging processes and criteria will vary according to intended scope and audience. Some awards have a "specialist" judging process, whereas other awards treat the judging process as a mechanism for a wider audience to learn about housing layout and design. The *Civic Trust Awards* encourages lay people, architects and local planners to learn from each other. The *Greenleaf Awards* encourages lay people in the sponsoring organisation to learn something about landscape. The Evening Standard invites members of the public to judge professionally selected schemes.

A relatively small number of organisations appear with some frequency across the 6 key awards:

- **Volume house builders**: Barratt, Crosby Homes, Countryside/Copthorn, David Wilson, Redrow, Persimmon/Beazer and Taylor Woodrow/Bryant.
- **Small house builders**: Ballymore Properties, Jones Homes, Octagon Developments, Wates, Beechcroft, English Courtyard Association, Environ County, London Buildings, Millwood Designer and Urban Splash.
- **Social housing providers**: the Guinness Trust, the Peabody Trust and Ujima Housing Association.

The notion of 'exemplar' schemes emerges as an interesting issue in its own right. Some of the exemplar schemes, as noted above, relate to rather narrow categories of housing design. These may, for instance, relate to 'energy saving' or 'best exterior design'. Others may relate to a much wider range of housing design considerations, typified by the *Housing Design Awards*.

If such exemplar schemes are to be useful to future practice, it may be appropriate to consider how the different components of housing design can be balanced in order to achieve the objectives associated with both improved housing design and urban renaissance objectives.

In this context, it is useful to explore the manner in which different types of house builder achieve this balance in terms of their awards competition behaviour, and whether particular trade-offs made are associated with different types of house builder. It would also be relevant to consider the manner in which different conjunctions of actors (developers/architects/planners) achieve different trade-offs in particular contexts. The selected case studies examined in subsequent chapters pursue these themes in more depth.

A great variety of housing schemes receive awards. People involved with the various awards feel that the quality of housing schemes has significantly improved over the last few years, and that designs that conform to the urban renaissance agenda are well represented. In contrast, however, there is a view that both good greenfield and sustainable schemes are poorly represented.

There still remains an issue concerning the role of design awards for housing in stimulating the very best, in comparison with the role of awards in achieving certain desired standards. A dichotomy is evident in both the criteria adopted in different awards, and in the orientation of different types of housebuilder towards different housing design awards. The consequences of this for the *Housing Design Awards* in particular are considered more fully in Chapter 6, where choices in the design of a housing design award are considered, as well as also in the conclusions and recommendations to this study.

Chapter 5

Do Awards for Housing Design Influence Industry Practice?

Chapter 5
Do Awards for Housing Design Influence Industry Practice?

Introduction

Interviews with developers, their architects, the local planning authority, marketing teams and a number of residents have allowed us to explore how awards are viewed by people who have been associated with thirteen award winning schemes. In addition, it has been possible to consider whether awards have influenced industry practices by encouraging better design in general. While the initial focus of our interviews was on the particularities of the case study selected, they proceeded to examine the wider ramifications of housing design awards in the pursuit of the urban renaissance agenda, and their relevance to the broader design and development practices of each of the actors. The interviews were reinforced by examination of five developers' recent portfolios of housing schemes. A list of persons interviewed is included in Appendix 3. Appendix 4 provides illustrated summaries of each case study.

To analyse the results an actor-by-actor approach has been adopted because developers, architects, local planning authorities and residents clearly have different interests at stake, and different roles to play, in the production of housing. To further refine the analysis, the developers have been discussed according to the categories used in the previous chapter – from large-scale volume developer to social providers. Discussion of each professional actor's assessment of the influence of housing design awards is extended to embrace their general attitude towards:

- the urban renaissance/sustainable communities agendas;
- achieving higher net residential densities;
- mixed use;
- mixed tenure;
- energy efficient buildings;
- site and context, and
- architectural design.

More specifically, questions considered the extent to which awards both acknowledged these issues, and if this acknowledgement in award giving encouraged the interviewees to pursue these qualities in future schemes.

The house builders

Although there are significant differences amongst and between different categories of house builder, the following general conclusions might be drawn from the case study interviews:

- The house building industry regards the *Housing Design Awards* as essentially an architectural award with a strong modernist/contemporary design bias. As a result they tend not to enter the award as they feel that this is contrary to the form of housing that they typically develop.
- They contrast this award with other key awards which they consider more industry-focused and which have a stronger base in consumer preferences and industry practice. Generally, they place more trust in the latter for these reasons.
- The industry generally has a relatively low awareness of design awards although this is definitely increasing. Some sectors of the market are very aware of design awards.
- The prospect of winning awards and the resulting exposure does not directly drive good design. The principal benefit of awards is seen as providing acknowledgement and kudos for those development teams whose efforts secured the award: motivating staff was seen as the major benefit of awards. Awards were also seen to be particularly important in attracting new, high quality staff who might be attracted to the award winning status of the company.
- By implication, and reinforced by direct comments from the developers, housing awards were not seen as strongly influencing the content or the quality of design outcomes across the bulk of housing production.
- PPG 3 and associated design advice were seen as much more important factors shaping the nature of housing production, even if there is a general absence of proactive design control within local planning authorities.
- The shift of development to more complex and constrained brownfield sites has resulted in more architectural patronage by the house building industry, and this has done more to raise design quality than any other factor.

Volume market house builders

The value of awards: The volume market house builders were the most sceptical of groups about the value of design awards generally, and particularly the more 'professionally-oriented' architect-dominated awards. The source of this scepticism is based largely on the view that volume house builders research and respond to consumer preference, and hence feel that they have a better appreciation of public preferences than do the juries of the 'professional' awards. They perceive tensions between the architectural profession's desire for contemporary design and its more modernist design agenda, and the industry's assessment that consumers are generally very conservative as expressed through their preference for traditional housing forms. This tension extends to other aspects of housing design. The volume builders place less importance on site planning and external design and more on the internal space and its layout. They also attach less importance to context-driven design, preferring to rely on a number of standard house types and finishes.

Volume house builders place more faith in the more industry facing awards to provide an affirmation of the quality of their products. It was commented that the *What House? Awards* and the *National Home Builder Design Awards* in particular seem to be more in tune with consumer concerns. These same awards are also more popular with the development industry because their results are effectively promoted to potential consumers through the national press. The *What House? Awards* are also associated with a publication that is popular amongst sales and marketing people in the house building companies, hence their positive disposition towards it. The *Greenleaf Awards* were also popular because they were relatively easy to win and significant numbers of awards could be garnered annually.

Award Entries: Generally the house builders leave it to their architects to enter the professional oriented awards. A number use specialist public relations firms both to determine the most likely awards and the schemes with the greatest potential, and then to promote the awards once they have been successful. This can dramatically improve their strike rate as the consultants carefully target the schemes and the entries to particular awards. Many firms are becoming increasingly conscious that awards can be important to company standing and reputation, and especially useful in advertising campaigns as a more objective measure of quality. They are beginning to take a more considered and strategic approach to award submissions and regional offices are carefully selecting entries to maximise their success. However, some companies consider that entering a high number of award competitions is expensive in terms of financial outlay and staff time. Others comment sardonically that the real expense is attending the awards dinner!

Attitudes to design: It is apparent that some of the volume house builders do not attach great importance to increasing design quality or in pursuing more sustainable patterns of building. However, on flagship schemes where market conditions or location might enable higher values to be achieved, or where the retention of external architects might obviate a planning permission or deliver a more marketable product, the volume builders will place a premium on design innovation and retain specialist, often award-winning architects. Where such schemes are successful in awards, this success will be used to emphasise the general quality of the house builder's product and the breadth of its appeal. Often in newspaper (and rarely television) advertising, award success is used as a badge to create an impression that investment in bespoke, quality design is the norm whether the design is traditional or contemporary in styling. Some companies noted a concentrated period of interest in their housing schemes following national exposure of their well-designed, award-winning schemes.

The volume builders appeared to be rather sceptical about using external architects to design anything but a small bespoke component of their production. Architects are perceived to have higher fees (20–30% higher was typically quoted), to slow down the design and delivery process, and to produce site plans which are expensive to build and with some houses that are difficult to sell. Further criticisms relate to architects' tendency (or perceived tendency) to use non-standard products, more expensive materials and create more variation and complexity in design, and these also add to costs. Architects do not give enough attention to the standard house types that are popular with buyers and many externally designed schemes have to be modified to take on board sales and marketing expertise.

These perceptions inevitably colour their attitudes towards those housing design awards that are seen as design and architecture rather than consumer focused. That said, there is a noticeable tendency for the volume builders to establish and expand in-house architectural teams to improve their site planning and urban design, but to do so within the constraints of traditional house building practices, standardisation mechanisms (generic house or apartment plans) and cost control. Most importantly, volume house builders consider that in most development cases the market will not bear the costs of better design. As one company expressed it:

"...there are many sites where the local market will not support significant investment in design quality."

In such instances, perhaps the vast majority of actual projects, an investment in design quality is perceived as an additional risk and one unjustified by their own market research. Certainly winning awards offers no apparent benefit to the house builder in such situations. Volume house builders argued that awards have virtually no direct impact upon the ease of marketing or the demand for particular developments. In fact only 'project awards' (awarded by *Housing Design Awards* and *National Home Builder Design Awards* for schemes that have not been built) would have the potential to directly influence sales. Even these, however, in the opinion of the few sales teams and marketing managers who had experienced them, did not do so to any appreciable extent (the time horizons are often too tight), though they might attract buyers to subsequent phases of large schemes. Marketing material will reflect housing awards in such instances, and it may help to increase and widen interest in the product, but it does not appear to produce any direct financial return.

The benefit of awards: A common response regarding the benefits to be gained from winning awards was that they enabled the commitment and professionalism of the development team to be acknowledged, and provided kudos external to the views of company management. They were seen as important to motivating development teams and maintaining staff morale in a highly competitive and hard working industry.

However, crucially there appears to be a limited transfer of lessons learned through award winning schemes to more 'standard' schemes within their portfolio. Both the nature of the industry and lack of internal feedback mechanisms in the design and development process, alongside a continued scepticism as to when design pays and when it does not, works against awards helping encourage the strengthening of the wider product. All interviewees indicated that entering awards has no direct impact on subsequent design schemes or practices operated within the company although they would select good schemes for award programmes if they felt that they conformed closely to the criteria or categories available.

In order to explore potential transfer of ideas from award winning schemes to the builders more standard product, the research analysed a range of marketing brochures for current projects. Two of the case studies are considered in more detail below, however in both cases it appears that carrying through commitment to design is fairly weak. Awards are seen as an end in themselves rather than a means to an end.

Developer 1: Typical volume builder

The company develops a very wide portfolio that includes bespoke developments on high value and high-density sites as well as a very large number of pattern book schemes. The London schemes tend to be very distinctive in comparison to regional schemes. The company seems to have commitment to original bespoke design where it is felt the market will support it but in other areas people are clearly buying a standardised product. Bespoke schemes vary stylistically, although only very specific schemes could be described as architecturally interesting. Commitment to urban design is difficult to judge but appears fairly weak, particularly on the standard schemes where house types do not appear to be markedly varied and houses are strung out along an access road. Some of the high density schemes are within gated tower blocks with little reference to their immediate context or attempt made to contribute to a pattern of streets or other public spaces. There is evidence of mixed tenure within the London schemes, although in other areas this is difficult to judge. Within certain locations schemes can be high density, but there is very little evidence of mixed-use. In general the company appears to demonstrate only partial commitment to design, and then only in certain locations. The company tends to use its success in design awards to create an impression of commitment to design quality across its portfolio, but which it can be argued is not evident in the bulk of its production.

Developer 2: Typical volume builder

The company has a standardised approach to housing development and design, evident from both their brochures and web sites. There is significant reliance upon specific, standardised floor plans and layout for the majority of the schemes reviewed, although many of the layouts appear to be well integrated with their context. There is some evidence of attempts at place making, although there is a very heavy reliance on standardised road forms. Certainly there is little evidence of innovation in house form, and the house plans tend to be differentiated using a standardised palette of materials and window/door forms which allow variation amongst familiar pastiche styles. There is some evidence of building apartments at higher densities but the vast majority of schemes appear to include mainly detached or semi-detached housing. No evidence of mixed uses was found, and only a small amount of mixing of tenures was noticed on particular schemes, although plans do not always clearly indicate this.

Volume bespoke house builders

The value of awards: The volume bespoke house builders share many of the same general attitudes towards housing design awards as their larger counterparts and demonstrate a preference for industry over professional oriented design awards. However winning awards is generally regarded more positively as they are seen as an important affirmation of the quality of their product – which compared to the larger 'market' builders, is an important differentiating feature of their schemes. Awards are particularly important to such companies and are seen as an endorsement of their quality product, whilst they also provide reassurance for customers of the quality of the company. This is regarded as important in marketing and sales as properties are often sold off-plan. Such developers make much more routine use of external architectural practices, and are more likely to enter the more professional awards.

Award entries: Volume bespoke house builders tend to rely upon the architects to enter the more professional oriented awards, but are generally more aware of the full range of industry awards and how to target them. Some have made a determined effort to scoop a wide range of awards. Others focus on more exclusive products because there are concerns that subsequent award publicity might damage their relations with their buyers and their desires for privacy.

Attitudes to design: A number of medium size and bespoke house building companies point out that they have specifically set out to differentiate themselves from volume builders by using good design and environmental sensitivity as a key component of their corporate identity. For many, the pursuit of design quality is a key aspect of their business because they are dealing with high value, attractive sites in sought-after locations. Equally, they may have large green and brownfield sites where an investment in design quality, particularly good master planning, is seen as the means of creating higher returns over the long term. These developers are particularly conscious of the ability of good design to overcome planning objections and, where local politics do not dictate otherwise, to maximise the value of the site.

All the companies interviewed recognised that the increase in consumer concern for design quality is acting as a real driver and promoting further opportunities for companies with design expertise and an innovative approach to development. Their schemes are often targeted at a younger audience who present a new generation of buyers with less traditional tastes and more prepared to consider modern designed schemes.

Much of the bespoke house builders' portfolio is at the 'luxury lifestyle' end of the market, and often companies utilise very high specifications and detailed design to carry the design concept and differentiate the product. This type of development tends to be designed for specific sites, and necessitates close working practices between architects and developers, and in particular many stressed the importance of good design in helping unlock the potential of difficult brownfield sites.

Only one company truly celebrated the potential of private architects to create original and practical designs, and has made this approach part of their company ethic. Architectural practices and consultancies are often retained, and there tends to be a symbiotic relationship between developers and architects in creating the overall design concept. Generally however, good architects within the housing/residential sector were regarded a scarce commodity, and at present the skills base tends to be confined to specific architects within particular practices. A number of companies either have, or are developing, in-house architectural expertise and using this to marry the best of the external ideas with their standard practices.

As with the large volume house builders, marketing and development brochures of schemes across the bespoke builders wider portfolios were analysed to determine the extent, if any, to which lessons learned through award winning schemes had been adopted in other schemes.

Developer 3: Larger scale bespoke house builder

The company tends to develop quite large schemes in phases and there is strong emphasis on a clear masterplan with reference to urban design, place making and townscape concepts. Housing layout is typically informal and pseudo-vernacular in styling, although there may be some more contemporary elements. The company develops a number of standard house types but will vary their plans for layout purposes, although areas of certain schemes appear to be bespoke. The company seems to be happy to develop to a high density if the market will support it. External treatment of façades seems to vary considerably and with varying degrees of success. The company seems to be happy to challenge accepted norms in highway design for the sake of realising a more imaginative scheme. There is evidence of mixed tenure elements within some schemes and some evidence of mixed uses. The approach to design varies across the portfolio, and some schemes certainly look more interesting than others.

Developer 4: Large scale developer with a focus on apartment buildings

A strong emphasis on apartment buildings means that all the schemes were bespoke. The range of contexts means that treatments are quite varied, ranging from very contemporary to pastiche, with a slight emphasis on the former, especially in the larger city centre blocks. There was some evidence of mixed uses, although in general blocks tend to be residential. A number of schemes were clearly on previously developed land. Most schemes appear to be sensitive to contexts, although being individual apartment buildings the schemes have not been required to create distinctive "place" characteristics.

Niche market house builders

Attitude to Awards: These companies tend to have already been converted to the design quality agenda and the awards that they win do occupy an important place in their Annual Reports and their marketing material. Despite this interest in awards however, it is perhaps unrealistic to suggest that awards themselves act as a key influence on development of the product. Rather, these companies have aligned themselves with the discriminating housing consumer, and are looking to be at the forefront of innovation in design, construction and community building.

Award entries: The smaller, niche builders typically rely on architects to make entries to the more professionally oriented awards such as the *Housing Design Awards*, however they do tend to get more involved in the selection and entry of projects for industry awards. A number of niche builders will not consider the *Housing Design Awards* because they perceive them to be very architecturally led with a bias towards innovative, contemporary schemes, although their architects have been known to persuade them otherwise.

The research revealed that many small companies feel excluded from the awards due to the entry requirement of five houses or more. Some very small companies can also find the entry requirements quite onerous. One interviewee commented that their company had experimented six months ago with an award and spent much effort on its entry, but that they have since received neither acknowledgement nor feedback, and so they felt discouraged from further entry.

Attitude to design: Smaller, niche market house builders tend to use good design as their unique selling point and have often developed long term relationships with innovative architectural practices. As noted by the bespoke volume builders, the value of using external architects is recognised in maximising development values and site densities as well as persuading local planning authorities that such projects add rather than harm local amenity.

Some of the companies interviewed specialise in finding brownfield sites where the design challenges are the greatest and where good design will have a positive impact upon the speed of planning decisions, the floor space achieved and the quality of housing environment created (for both residents and neighbours). Others have marked out contemporary design, environmental sensitivity or building innovation (or indeed all three) as factors capable of attracting particular consumers and of delivering innovations that will give the company a competitive edge. They see consumers becoming more discriminating and government policy setting greater challenges, and they want to position themselves to take full advantage of these trends. As with the other house builder categories, investigation of the product portfolio of awards winners helps indicate whether their award winning schemes are reflected of wider company practice.

Developer 5: Top of the market medium sized bespoke house builder

There was no repetition in this portfolio of schemes, with each project individually designed. Schemes seem to be either contemporary or vernacular in form and styling and the company seem to be happy with either approach. Densities vary, although there is plenty of evidence of higher density schemes and apartments, especially within London. Standard detached and semi-detached housing schemes are very scarce.

Developer 6: Small developer of contemporary apartments

This developer produces site specific, architect designed, bespoke apartment schemes on low value brownfield sites in strategic locations typically on the edge of city centres. Each scheme is therefore completely original and in a modern idiom. Schemes are typically relatively affordable, and always include a mixed use element in the form of either live/work units, office space, bars or shops. Housing is typically at a high density with parking at a minimum standard and sometimes underground. Apartments are open plan, with a strong emphasis on the qualities of both space and light. Buildings typically form part of a wider urban renewal scheme, and the company puts a lot of emphasis on the quality of the public realm around its dwellings.

Social housing providers

Attitudes to Awards: The social housing providers interviewed demonstrate quite different attitudes to awards compared to their private house builders counterparts. They do, however, share the view that awards have value in both recognising design achievements and motivating staff. Whilst award winning schemes often feature prominently in housing associations' Annual Reports, some associations are wary of promoting award winning schemes for a number of reasons. Concern was voiced that award success may attract the attention of auditors (*good design must cost more money*), prove less attractive to prospective tenants, or prove problematic in terms of both running costs and long term maintenance and management. The housing associations were more interested in good design from the perspective of life cycle costing, ease of management, and the effects of higher level of specification on both.

Entries to Awards: It is rare for a housing association to actively pursue any award for housing design. The prevalence of some social housing providers in the *Civic Trust* and *Housing Design Awards*, for example, does not therefore clearly reflect the interests of the sector as a whole, but instead of their architects who submit the schemes for the awards.

Attitudes to design: The associations interviewed appeared to have less resistance to contemporary design than private developers. However they did recognise the need to conform to local private market norms and not to be distinct from the private components of a development (many of their housing schemes are section 106 components of a private development). This extends to providing loft-style apartments in certain central urban locations and adopting more contemporary design styles in higher density schemes.

Some housing associations are more experimental by necessity, particularly in more expensive inner city locations, because in order to obtain sites they are forced into partnership with a variety of public bodies, institutions, businesses and developers. Such projects by their very nature have mixed uses and frequently mixed tenures, adopt low car parking standards and have high densities. The projects benefit from much more investment in design and innovation and they are very much in tune with the key objectives of the urban renaissance and sustainable communities. However, if land prices continue to escalate housing associations fear they will be priced out of the major inner city markets altogether in the future.

Architects

Attitudes to Awards: Architects specialising in housing design are a comparatively small group. The interview sample included some notable architect developers, established practices with a particular niche (for example modern apartment designs), and younger practices seeking to establish repeat commissions and new business. The professional design awards are clearly a major target for these practices, although this is balanced by scepticism as to whether anyone pays much attention to these awards, and especially whether the developers attach much weight to them. Nonetheless, they see awards as positive not just for peer esteem, but also for the growth of their business.

Architects regard the *Housing Design Awards* as the most professionally prestigious, and of all the actors involved in the design and development process of a given scheme, it tends to be the architect who will take the lead in entering schemes for this award. They consider that winning such an award can cement a relationship with a developer and that this has lead to greater penetration of their practices into the mass housing market. Most of the practices involved in the research had been involved in subsequent housing schemes with house builders whom they had previously collaborated with on an award-winning scheme.

Architects prefer the *Housing Design Awards* to any industry award as they are regarded as having an architectural focus, are perceived to be more oriented to design innovation, and identify the more unusual schemes that tackle difficult sites. A common assumption is that the awards are aligned predominantly with one of the four sponsors, with many regarding the *Housing Design Awards* as exclusively an RIBA initiative. The variety of schemes winning the *Housing Design Awards* in terms of both regional location and size is also appreciated, especially by young architects working outside London who are keen to raise their national profile.

Entries to Awards: Architects tend to assume the responsibility for entering the professionally oriented competitions and preparing the design sheets necessary to meet any requirements. Much effort goes into these entries with text, graphics and careful choice of illustrative material, and there is some resentment that the client often shows little interest in the entry and does not contribute to the costs. Architects have a rather less enthusiastic attitude towards the industry-focused awards, regarding them primarily for the house builders themselves. They do recognise however that the more commercial recognition is perhaps more likely to generate new business.

Attitudes to design: The architects interviewed felt that prize-winning schemes do generate repeat business, but the effects are rarely dramatic and new business has often proved quite slow to develop. Some of the more innovative house builders do keep a close eye on winning schemes and invite the practices in to give presentations, but it may be some years before a scheme comes up where they could be involved. Architects are glad of the plaudits from government and the professions, and the publicity and recognition it generates, and they applaud the work being done by ODPM and CABE to promote the case for more innovative, contemporary and sustainable design. This support is very important to their morale, especially as they seek to attract a steady flow of commissions.

Architects argue that they are being increasingly employed to design housing schemes because house builders recognise that only their skills can get the best out of a difficult to develop, brownfield site that requires a bespoke, high density design solution or a scheme at the top end of the market. On a number of occasions they have been approached as a result of difficult or failed planning negotiations. A number of the architects interviewed indicated that they were less keen to work on standard greenfield housing schemes as they would quickly lose interest in the work.

Some architects expressed concern about the tight time scales required by house builders and pointed out that this work is less profitable than that for other clients. One architect indicated that they were paid £100,000 for securing a difficult planning permission on a £12 million project, and they were given six weeks to achieve it. Architects also indicated that the technologies and construction techniques being used to realise higher densities and higher values are often quite sophisticated (concrete and steel frames, extensive glazing systems, complex servicing arrangements) and that this is not always understood by the developers who will ultimately benefit from them.

Some practices bemoaned the fact that developers were free to change even prize winning schemes and those with detailed planning permission, leaving the architects out of decisions on what actually got built, and putting at risk the trust built up between designers and local planning authorities. One practice noted the disinterest of many architects in project supervision and management and considered that the profession had only themselves to blame for being treated in this way. However, since the practice concerned had a commitment to project management they were all the more disappointed to find themselves not only unable to implement their scheme, but having to watch it stripped of some of its design quality.

Architect-developers were able to exercise all their design and project management skills and were also free to find sites and development partners. The altruism of some of these companies was very refreshing and their design achievements impressive, resulting for example in some very interesting collaborations with registered social landlords. Some had made significant positive impacts upon particular neighbourhoods and localities. There was evidence of a ploughing back of potential development profit into a larger social housing component which allowed them both to attract local authority subsidies for affordable housing and win planning permissions more effectively. Perversely, perhaps, such companies suggested they were less interested in housing design awards because they do not need them to attract commissions. However, awards undoubtedly help their dealings with development partners and local authorities, and are an affirmation of their professionalism and commitment.

Local planning authorities

Attitudes to awards: Planners noted that they were proud to be associated with award winning schemes and that this enhanced their job satisfaction. The winning schemes themselves helped to provide benchmarks of what could be achieved locally and these were used as exemplars for other developers. There was some cynicism about architectural fashions and the fickle nature of public and professional taste that could lead to prize-winning schemes being ridiculed by the next generation of critics and amenity spokespersons, but this seems a rather dated view.

Attitudes to design: Winning awards was not a prime concern of development control officers, although adding value to applications might have been expected to have been given more emphasis than it was. Many planners adopted a rather neutral attitude to design quality, suggesting that their role was to prevent poor design. As one planner expressed it: '*the test is not is it the best? The test is, is it good enough?*

Nevertheless, planners can still exert significant influence on design quality, particularly where conservation or landscape designations impact on sites, or where local residents or councillors express concern about development impacts. A number of the award winning schemes had come about as a result of planners challenging the quality of previous schemes submitted by the same companies. In addition, the innovative design features of schemes sometimes resulted from attention to and enforcement of planning requirements.

Views from other actors in the design and development process noted considerable variation in local planning authority approaches to design. They commented that although there were some authorities that were regarded as beacons of good practice, many are under resourced, are unable to retain high quality staff, and tend to be reactive rather than use proactive design policy and guidance. Development control officers in particular are hampered by high workloads, and have varying levels of design awareness or skill. A particular concern was the refusal by some planning authorities to enter into either pre- or post-application negotiations because of pressure to make decisions within eight weeks. As a result, both developers and architects commented that many of the positive interventions that planners used to make to improve the quality of schemes are being lost.

Resident responses

House Consumers

Marketing experts suggest that residents are interested in design awards largely to the extent that they confirm their own judgements about design. If they therefore do not like a scheme, the fact that is has won a design award will not have any impact on their decision to purchase. Awards also do not appear to impact on selling prices, but they will have some impact on the confidence that consumers will have in a company's general brand. This is particularly important where consumers are buying off a plan, but can act as a focus for disappointment with the product where initial aspirations are not matched:

> *"My wife, my son and I have purchased 623 and 421 as investment vehicles intending to own and let them long term. The properties were chosen for location, price and quality. [The company] have a good reputation and the specification was good, but I regard the whole design award business as a gimmick. It seems to me that if an architect receives an award for a building and that award is subsequently used as a marketing tool then that architect has a responsibility to ensure that the building lives up to the award. In the case of [scheme] this has not been the case. The finished object falls considerably below the standards that I had expected, due mainly to [the company] failing to ensure that adequate quality control was employed during the construction and finishing. Possibly in such a case the award should be revoked."*

Whilst the fact that a scheme had won an award had little influence on consumer behaviour in terms of purchase, many were nevertheless proud of the fact that they live in an award winning scheme:

"I was told that the apartments had won a design award when I went to view my property but it did not really affect my decision to purchase. I would have purchased anyway. The design quality and location were key factors in my decision to purchase. I liked the way in which space was maximised through the open plan design and I loved the fact that it was right in the city centre. I was also surprised at the price for which the property was offered. I am very proud of the fact that my apartment has won a design award and take every opportunity to mention in to people I know."

Councillors and Neighbouring Residents

Although councillors and neighbouring residents were not interviewed as part of this work, a number of other interviewees did comment on the views expressed by these groups on award winning schemes. It was often the case that neighbouring residents were very sceptical about any new scheme within their vicinity, and they would protest against any scheme that they felt would undermine the character of their area or which might introduce too much additional demand for parking on neighbouring streets. Commitment to raising housing densities is, in particular, often challenged at the local level where councillors and residents feel that even well designed schemes are inappropriate.

Objections become particularly pronounced however, where matters of style are being considered, and the badging of schemes as 'award winners' will often further fuel debate. Neighbouring residents may reject award judgements where they feel that a scheme is 'too modern', even if the intervention of either the Royal Fine Arts Commission or English Heritage has influenced Councillors to approve a scheme. This was found to be the case with Century Court in Cheltenham where the clash between the contrasting views of lay and professional tastes became clearly apparent:

Let's camouflage Century Court

MADAM – I write in response to Tom Gardner's letter (Letters, July 7) regarding Century Court and speak out for the 'silent majority' who see through the King's invisible clothes of this building.

The building is a triumph of the illustrator's and designer's art. It is a triumph of 'spin' over good sense. Designers and promoters

of Cotswold pastiche of Dutch and classical. It is endless. As is the pastiche of Century Court's attempted revival of modernism.

It is arrogant of architects to think that 'their' buildings can stand on their own. The surroundings are what constitutes the environment, the sense of place.

Unlike Bath, Cheltenham only

Who is to blame for ugly flats?

MADAM – I totally agree with the comments by Echo correspondents that Century Court in Bath Road leaves an awful lot to be desired in its design and lack of sympathy in architectural terms with Cheltenham.

However, rather than blame council officers, surely the ultimate responsibility lies with the

Extracts from the Gloucestershire Echo referring to the Century Court scheme

The example of Century Court highlights how the innovation evident in and celebrated by national award winners can be regarded as an imposition on local character. How criteria and award procedures might address these concerns is considered in subsequent chapters.

Conclusions

Awards would appear to be only a strong motivating force for architects, and they focus largely upon the *Housing Design Awards* as their premier professional accolade, with the *National Home Builder Design Awards* as a second best. House builders are typically not strongly motivated by design awards, and instead tend to prefer those that they perceive to be more reflective of industry aspirations and consumer preferences rather than those which are more professionally oriented. Nevertheless a number of house building companies, and particularly some of the more niche or bespoke companies, do take greater interest in awards and acknowledge their potential benefits. They are much better informed about both awards and the potential contribution that external architects can make to maximising development value and delivering schemes that can get planning permission rapidly. Even with more proactive companies however, awards do not appear to greatly influence their day-to-day practices.

These generalisations must be tempered by three observations. Firstly, companies are becoming much better informed about design awards, are beginning to monitor their award achievements carefully, and to compare themselves against other companies. For example, one volume house builder deliberately set out to emulate another's achievement of a dozen *Greenleaf Awards*, and succeeded in doing so. Secondly, volume house builders are realising the value of a number of prize-winning schemes that can grace their national and local advertising campaigns, despite a reluctance to raise design standards significantly across the board. Thirdly, even house building companies who have no desire to employ external architects are reinforcing their own design teams and upgrading their design skills in-house.

A number of interviewees commented that a design renaissance is slowly taking place, perhaps less in response to housing awards than to a new generation of households with wider horizons and less constrained design tastes. They acknowledged an explosion of interest in interior design, garden design and more recently housing design as elements of lifestyle. As such, a property owning democracy is becoming property obsessed, especially as alternative investments falter.

The interviewees felt that more focused and positive planning policy since the mid 1990s has also contributed to encouraging shifts within residential design. In particular, the increased focus on brownfield development has made standard design solutions much more difficult to achieve (particularly after the first few crude attempts in a locality). Developers have recognised that only good design can achieve the three necessary requirements: to maximise development value, negotiate a planning permission and create a saleable product that will constitute a good housing environment. In addition, and despite the widespread under-achievements of design control, design sensitive development control continues to play an important role in raising design standards in many local authorities. A number of the award winning schemes resulted initially from interventions by planners typically rejecting inferior schemes and encouraging developers to 'go back to the drawing board'.

To be effective in improving design quality, awards for housing design have to encourage architectural patronage. There is little evidence that they do this directly but some evidence that they do so indirectly. Awards have their greatest impact where they encourage architectural practices to fully exercise their skill and ingenuity. Furthermore design awards bring architects important peer recognition and national publicity to bolster their image and marketability.

At the heart of the pursuit of design quality is the question of whether the market supports, or demands, a high standard of design. There is evidence that a number of companies use design quality as their unique selling point. They tend to be the medium and small bespoke/ niche builders, although a number of the large-scale volume builders have recognised the importance of design, particularly at the high end of the market or where they are seeking to attract more adventurous homebuyers or investors. Crucially however, as noted by one of Britain's biggest house builders, the majority of sites with local market conditions will not support significant investment in design quality. Ultimately it is this that ensures that the impact of design awards and associated initiatives on the large majority of housing output will be limited.

Chapter 6

Choices in the Design of a Housing Design Award

Chapter 6
Choices in the Design of a Housing Design Award

This study has explored how particular development and consumer groups respond to the design agenda for housing and more specifically existing but varied awards for housing design in England. As a result it is now possible to indicate how design awards could be created to respond positively to these groups and their interests. In so doing tensions are also inevitably raised: by making an award more attractive to one group it may well mean that it becomes less interesting to others. Equally, whilst an award's audience base may be widened through altering its focus, the value of the awards to its original core group may become somewhat diluted and thus lose meaning and influence as a result. In this chapter the main interests of the professional and consumer groups are brought into focus, and then the choices for how design awards might be formulated to respond to their interests are highlighted and discussed.

What stakeholders want from awards

Private developers

Although the types of private developer as discussed in previous chapters vary greatly in size, their interests are largely similar and will be discussed here as one group. They want a design award to:

- have integrity;
- reflect consumer concerns in its judging;
- be disseminated in a form that will be seen by their consumers.

Developers tend to enter awards that they feel they have a chance of winning, and they will target categories that match their output. They are not necessarily interested in innovation for its own sake. Importantly they seem to feel able to ignore awards that do not conform to these criteria. That said, bespoke and niche housing developers will often participate in awards that have an architectural focus because of the status enjoyed by such awards amongst their informed clientele. It tends to be their architects, rather than the developer themselves however, that enter schemes for such awards.

Registered Social Landlords

With the notable exception of a few providers, RSLs do not tend to show significant interest in design awards, and their interests seem to be adequately represented by their architects in the *Housing Design Awards* and *Civic Trust Awards*. Even those RSLs, where design quality is highly valued and who indeed often appear as award winners, are a little cautious about the value of awards to themselves.

Architects

The research confirmed that architects' acknowledge and recognise the benefit of professional peer endorsement for their work, and so they are keen that awards should involve judging informed by an appreciation of their design objectives. In particular, they wish to see innovation in both form and style celebrated, and so original and exemplary responses to established typologies (say terraced housing) should be award winners rather then high quality pastiche schemes designed in a traditional idiom. Architects want other architects to see that they have won awards, but they are also keen that housing developers and the general public should see that they can design high quality housing. In particular, architects think that award-winning schemes can be used to highlight the benefits of good design in the public interest.

Planners

Planners want awards that will raise the general quality of housing being developed within their areas. Although they are delighted to receive exemplary schemes, their main concern is to raise the overall standard of residential schemes. The fact that awards encourage architects to be involved in housing design is therefore important, as this is felt to raise design quality, but also general consumer expectations. Additionally, planners are keen to find suitable precedents that they can use in negotiations with developers to show what can be achieved in similar contexts. Local awards seem to be particularly important as many of the national awards are felt to include precedents that are not always particularly relevant.

House consumers

People who buy an award winning housing scheme are typically not that concerned by its award winning status, unless they feel that it endorses their own view of the scheme. Certainly any deference is short lived following the experience of living within schemes, where a particular concern for build quality seems to prevail over any other judgement. It does not matter how innovative or beautiful a scheme might be if the doors will not shut or the roof leaks. Marketing people feel strongly, however, that awards give consumers confidence that the quality of their product is good, even though the consumers will not be aware or concerned about the specific areas of judging or the criteria used.

Councillors and neighbouring residents

Through interviews with the wide range of actors involved, councillors and neighbouring residents were perceived to have a specific but quite far-reaching impact on the quality and form of residential development. In particular, they are seen to typically dissuade or oppose the development of housing in a modern idiom, and are concerned to keep densities low and car parking standards high so that the impacts of developments are not felt too heavily within the context of schemes. Competitions that endorse, for example, high density, modern schemes that are felt to be inappropriate for their context risk simply being dismissed as elitist.

The wider public

Beyond the concerns of specific consumers, there is a wider public who, according to interviews held with representatives from the media, show a particular interest in housing awards and award winning schemes. Their interest in awards is largely aspirational, although they also seek to be entertained by stories surrounding the people involved in award-winning schemes. To this end the media are keen that dissemination should be focused on human-interest stories as much as on the designs themselves. The public wants to be entertained and excited by innovative designs and they want to know, for example, how much it costs and what it's like to live there. Some journalists point out that in general the public respect architects and their judgements and so are not bothered if architects are, for example, judging other architects, and that award criteria do not necessarily reflect consumer concerns. Such coverage tends to be positive, although award-winning schemes that are dreadful to live in have equal journalistic appeal.

Choices in the formulation of design awards

The above discussion has tried to highlight what particular interest groups want from design awards. Using this analysis, the choices available in the formulation of design awards will be discussed, highlighting tensions that may arise when choosing particular approaches and seeking to satisfying particular audiences.

Judging procedures

As discussed in Chapter 2, previous research has identified the tensions often apparent between lay and professional concerns in housing design. This research has highlighted however that lay concerns appear to be fragmented, and that as the above discussion indicates, it is possible to broadly distinguish between consumers, neighbouring residents and the more general public, all of whom want something slightly different from the award giving process. Despite this, awards need to consider how they balance a concern for lay interests which particularly pre-occupy both consumers and developers, against the professional or 'public interest' agendas of Government, planners and to some degree architects.

Many awards have responded to this procedurally by including a lay element into the judging process so that schemes selected can be seen to match public preferences more closely. The experiences of the *Civic Trust, Evening Standard* or *Greenleaf Awards* offer models that might be considered. The *Civic Trust Awards* have a lay person on the judging panel of each scheme who will research public opinion locally and provide comments to the judging process. The *Evening Standard Awards* have lay judges selected from readers of the sponsor paper, and the *Greenleaf Awards* use lay judges, but they adopt simple criteria provided by the professionals.

Judging criteria

Criteria tend to be fairly explicit for a number of design awards, however the extent to which they again reflect professional versus consumer concerns seems to vary. It is difficult to actually judge whether consumers and neighbouring residents are really concerned by the minutiae of such criteria since their interest in awards is not that strong. The interviews did provide evidence to suggest that innovations celebrated by professionals do not always lead to quicker sales. Encouraging sales is, however, a key factor shaping the involvement of house builders in particular awards programmes. The popularity of the *What House? Awards* in particular is shaped by the perception that the judging criteria focus on consumer issues, although the awards documentation does not really provide much information about how this is achieved or what criteria are adopted. The criteria for "Best Development" are rather vague:

> ***"Judges will be looking for a stylish mix of original designs, top quality site layout and landscaping, and variety of external elevations and plenty of design flair inside and out."***

Nasar's (2000) research on consumer preferences highlights how, in judging housing quality, lay people are concerned with matters such as visual richness, order, openness, naturalness, upkeep, historical significance, and liveability. Such criteria might be used to supplement professional urban design criteria in certain award programmes in order to reflect lay concerns more. Such expansive lists of explicit criteria may become difficult to use, although they would inform and probably help 'soften' professional judgements. Were this to happen, however, it is likely that architects might become less interested in any such awards. Architects view the opinions of their architectural peers very highly when matters of design quality are being judged, and there may be a potential fear that increased lay involvement may lead to a 'watering down' of the awards' value. Equally, it is also unclear whether developers would actually become more interested in such programmes if an increased focus on consumer views was provided. Despite assertions that their involvement in awards is geared towards reflecting this consumer interest, their involvement tends to also be dictated by matching the involvement of their competitors. Similarly, the awards selected for entry tend to be those with a large number of categories which make them easier to win.

Types of award and judging categories

Award types and judging categories can be seen to reflect the pre-occupations of the award givers, but at the same time they act as an incentive for certain groups to enter particular programmes. Our earlier discussions highlight how award programmes can contain exemplar awards, benchmark awards, category awards and also awards for outright winners. It also notes that exemplar awards are particularly attractive to architects who seek professional kudos and recognition from innovative designs. The benchmark awards are attractive to developers, as success is more certain and hence the effort is worthwhile. Both developers and designers value the category awards, especially where categories are aligned to particular products that are being designed, and for developers especially where more than one award is given in each category. Finally, being an outright winner is highly prized and particularly valued by the media.

Typically award programmes contain a combination of the above, although a key distinction can be drawn between those which have a large number of categories and award mechanisms providing a better chance of being successful on the one hand and those that only really celebrate exemplary schemes on the other. The former includes the *What House? Awards* which in 2002 had 72 chances of winning something and the *Greenleaf Awards* which in 2001 awarded prizes to 78 schemes with an additional 6 special category awards. The latter include the *Civic Trust Awards* which in 2002 gave awards to 17 exemplary housing schemes, and the *Housing Design Awards* that in the same year also gave out only 19 awards and commendations. Interviews with house builders illustrated the need to develop categories that reflect their product range (rather than the design agenda of the award promoters), and it needs to be easy for them to gain something that they can market to consumers. If a key aim is to involve architects, then the awards should retain greater 'exclusivity' and celebrate a small number of exemplary schemes.

Planners, interestingly, probably like the combination of award types that exist. They want to see developers entering awards and therefore maintaining their average development quality, but at the same time they will value exemplary schemes that illustrates potential future trends nationally, and maintains the interest of architects in housing design.

Although all award programmes will include categories that reflect the urban renaissance agenda, the structure of some categories in particular awards can discriminate against attention to it. The *Housing Design Awards* is distinctive as it essentially judges all schemes according to the same criteria, with circumstances surrounding the context providing the basis for variation between schemes. In contrast award categories for, for example starter homes, judge only the merits of the individual house, even if its located in a poorly designed scheme and without reference to the more strategic concerns that urban renaissance criteria aim to reflect. Again, house builders' interest in programmes like the *What House? Awards* probably reflects the fact that they can design individual houses and win awards without wider context/ neighbourhood issues being taken into consideration. Award programmes do, however, often contain categories for development which explicitly relate to urban renaissance concerns, such as development on brownfield land or re-use of an existing building. Many of the case-studies interviewees in this research also reflected that developments are increasingly complying with urban renaissance concerns. This, however, was typically because of the need to comply with planning requirements and the fact that brownfield sites tend to require more innovative and interesting design solutions.

Dissemination

Awards are publicised:

- using trophies, plaques and flags located in offices and on site;
- through exhibitions;
- profile in both regional and national newspapers;
- through special publications and supplements;
- using web sites.

House builders in particular regard award dissemination as a central reason for participating in awards, and they are very keen for their consumers to see their success, hence their interest in awards that are effectively disseminated through either the national or regional press. Typically the interviews highlighted that developers are not marketing their award winning schemes as they are normally already sold, but they will use the award to support their brand, whilst they might also use the media profile to advertise other schemes in the vicinity. Without such a high media presence house builders will again be less attracted to a given award programme, which explains in large part their interest in being involved with the *National Home Builder Design Awards*, the *What House? Awards* and the *Greenleaf Awards*.

Architects also welcome high profile marketing of their success, although they tend to be more involved in the *Housing Design Awards* that do not gain such high profile within popular media coverage. Probably this explains why interviewees highlighted that new business with other developers often seems to result more from word of mouth, as developers in particular are not necessarily following the *Housing Design Awards* that closely.

Newspaper advertising from Barratt highlighting their success in competitions

Different forms of dissemination also seem to inform developers, architects and planners about housing design potential in different ways. The fact that *What House? Magazine* and *Show House* are popular magazines for private house builders means that house builders have confidence in the *What House? Awards* and perceive it to be in tune with consumer aspirations, whilst they also seem to be largely ignorant of the *Housing Design Awards* documents. By contrast, architects and planners appear to make more effort to seek out exemplary schemes and new ideas and hence they will use the *Housing Design Awards* brochure and the associated *Pick of the Crop* publication.

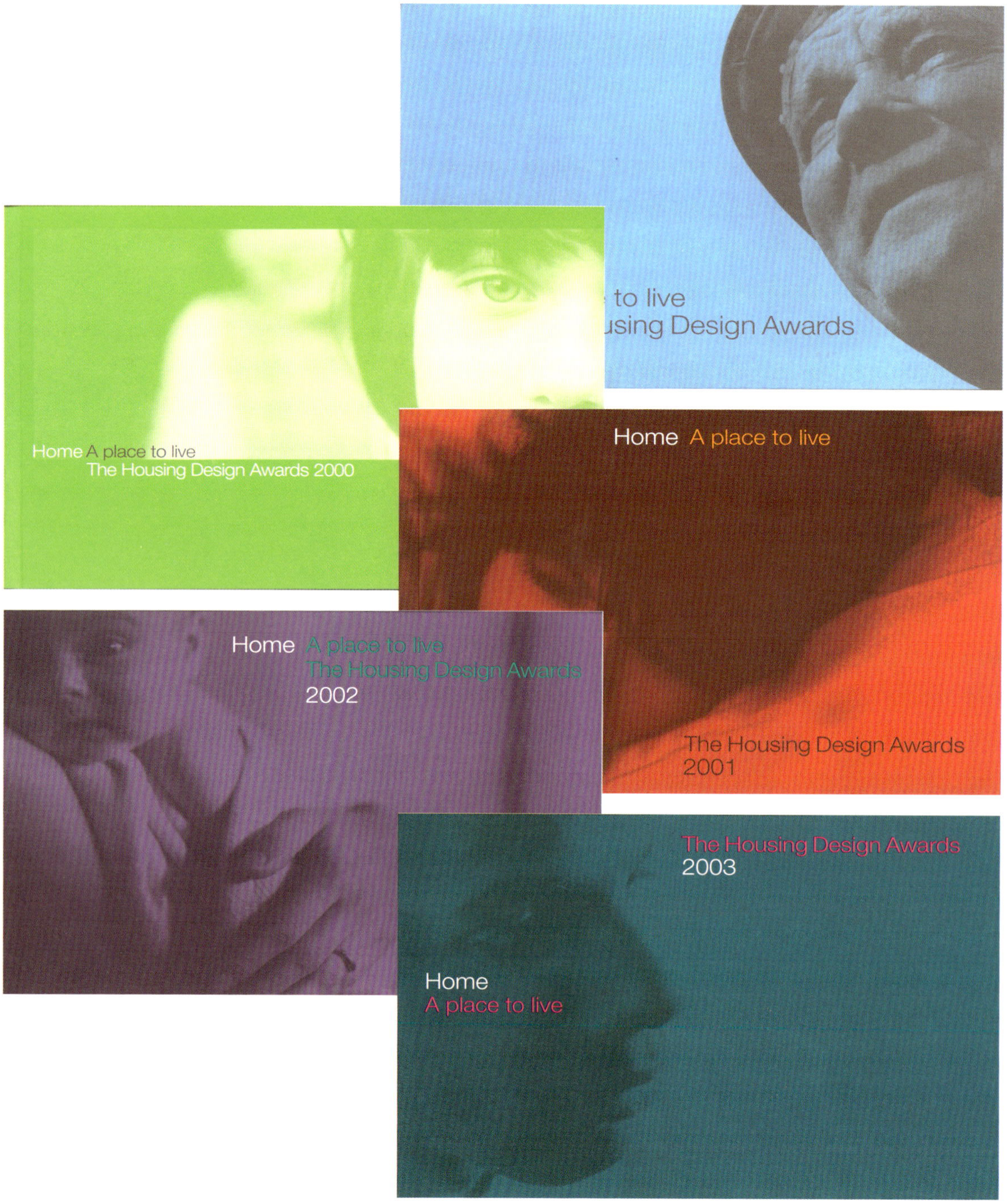

Architects and planners make good use of the Housing Design Awards brochures

Conclusion

The various actors in the process of producing and consuming housing do not share a common view of awards that would allow consensus to emerge as to how they could best be formulated to serve all needs and interests. Contrasts between professional and lay concerns in the evaluation of housing have been highlighted through the review of academic literature, and certainly developers are keen to suggest that they are interested in awards that reflect consumer concerns in particular. This analysis has also highlighted that house builders want to enter awards that firmly reflect their development concerns and product types; they also want the awards to be easy to win and widely disseminated. Whilst architects share interest in lay concerns, they are also clearly motivated by the potential status of winning elite awards that focus on design innovation and originality. In addition, they are also keen to seek out information about other exemplary schemes, hence their interest in publications associated with the *Housing Design Awards* in particular. Other groups demonstrate rather more ambivalence towards awards generally. Planners are happy with any mechanism that raises the general quality of design, and consumers only really engage in awards to the extent that they might confirm their view about the quality of a developer, rather than the quality of a specific scheme.

It can be suggested that the particular emphases seen within the key housing design awards considered in this research – whether towards the developer or designer end of the spectrum – fulfil particular needs and appeal to different audience profiles. There are clearly strengths in recognising and focusing upon a more niche audience rather than attempt to appeal across too broad a group. Nevertheless, most awards recognise the value of combining features to reach a perceived ideal mix. In the final part of this report, recommendations for the *Housing Design Awards* will provide details of possible options to address questions of audience, appeal and dissemination.

Chapter 7

Conclusions and Recommendations

Chapter 7
Conclusions and Recommendations

A key objective of this research has been to consider how the *Housing Design Awards* could be developed further to encourage better design of housing in England in line with the Government's policy objectives. The findings from the empirical work and analysis will now be drawn together to highlight our key conclusions and the options that are available for the future of the *Housing Design Awards* programme.

Awards for housing design can only be appreciated within the context of other mechanisms available to Government to control, guide and encourage better forms of development. Awards are distinctive as they represent a reward for a completed initiative, draw attention to good practice, and provide a media friendly mechanism to enable that good practice to be more widely disseminated to an interested audience. Despite these positive attributes, this research has highlighted the limited influence awards have in encouraging better housing design, particularly by the larger volume developers. Comparatively Planning Policy Guidance (PPG), enacted through the planning system, is regarded as very influential. Award winners often acknowledged the impact that PPG3 in particular has had on certain developments.

Awards, however, do have a very specific and positive role to play and the *Housing Design Awards* have been particularly successful throughout their history. They have a number of specific strengths:

- They are regarded, by both developers and architects, as the most prestigious national awards.
- They are most open to contemporary and modern designs, yet equally reward good design in what may be considered more traditional or vernacular styles.
- They aim to combine industry, professional and government views about what might constitute good residential design.
- The potential of winning them is a clear incentive to architects to maintain an interest in housing design and can provide a critical mechanism for getting them noticed by housing developers. These particular architectural practices are making a major contribution to better design in parts of the industry, and particularly when working with bespoke developers.

The awards inevitably also have their weaknesses:

- A commonly held perception – often reinforced through media coverage – is that the *Housing Design Awards* are the 'RIBA' sponsored awards. This selective association, common to both architects and developers, reinforces perceptions that it is a professional focused award.

- Partly as a result of this, and partly because of a perceived emphasis on contemporary design, the *Housing Design Awards* are also given an elitist tag.
- Dissemination is felt to be limited. In particular it is felt to be targeted at the design community, whilst information about winners does not reach the general public in a form that is of interest.
- It is relatively difficult to win.
- Volume house builders, not overly concerned with building exemplar schemes, may feel the effort of entry cannot justify the potentially low odds of return. Other awards, easier to win and perceived to be more in tune with the demands of 'their' market can provide a badge of recognition for the brand. Bolstering the brand, rather than concentration on the particularities of a specific scheme is perhaps more important. As a result it has a limited impact on the quality of mass housing that is produced.
- It is perceived to largely ignore lay tastes or seen as having an agenda to change them. This is because of the interest in rewarding innovative forms of housing and schemes built to a higher density on brownfield land. Such forms are felt by house builders to be contrary to most consumer aspirations.

As a result of this research a series of options are presented below as to how the *Housing Design Awards* might be changed to overcome some of these weaknesses and give them an altered focus. This reflects discussions in the previous chapter where it was argued that the way particular features of the award are designed might, in particular, encourage or discourage participation by particular groups. Providing a series of watertight options does not really reflect the diverse range of choices that might be made in the planning of an award programme, but it will allow general decisions to be made about the direction of the programme.

Options for the Housing Design Awards

Minimum change

The *Housing Design Awards* could continue to maintain their current focus. This includes:

- acting as a means to demonstrate best practice aligned to Government policy;
- running separate completed scheme and project awards;
- having relatively few specific categories;
- using professional criteria and judging all schemes equally according to those criteria;
- maintaining commitment to a belief that good contemporary design can be demonstrated through both modern and more vernacular forms and styles;
- rewarding a small number of exemplary schemes;
- employing professionals to be the judges;
- disseminating the results essentially to a professional or well informed audience.

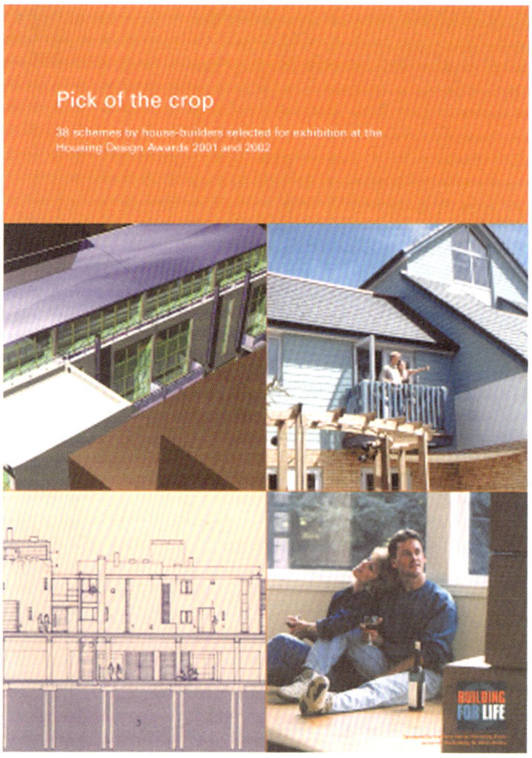

The Pick of the Crop publication should be maintained

The entry requirements and format work well and require no change. The awards process is considered streamlined and efficient, and changes to the awards as they are currently designed would have major resource implications. The current dissemination mechanisms used by the *Housing Design Awards*, including the web-site and award brochure, are also well received by the professionals and could be maintained in their current form. The *Pick of the Crop* publication, produced by the House Builders Federation, is also popular, particularly in its provision of information about precedents that are of professional interest.

Such an award would maintain the interest of architects in particular, whilst it would also provide clear practical examples for developers of the forms of housing that the Government wishes to encourage. There is clear evidence that PPG3 is having an impact on the forms of housing being developed, and such an award would continue to endorse exemplary precedents. By taking a broad, inclusive view of quality modern design, it would also continue to challenge developer views that consumers will only consider living in vernacular housing forms. When viewed within the context of other design awards, there is a strong case for maintaining the *status quo* for the award. A possible consequence of losing this focus would be that this award programme would simply become more like the others in its focus and form. This may provide access to a 'broader' audience', but the Housing Design Awards professional recognition is at risk of dilution. Of course maintaining the status quo will mean that the award remains largely regarded as elitist and as such of limited attraction to a wider number of housing developers. It can be argued that the strength and clarity of the awards are both its strength and weakness.

If the 'minimal change' option was to be pursued, there are a number of small additional actions that can be considered and which would not dilute of the award's intended focus.

- A single named winner might be considered appropriate. This would be a positive step to gaining increased media profile. Having one winner would promote controversy and debate and would therefore be more readily picked up and used by the media. However, there has been some debate as to whether having one winner would detract from the great variety of housing being considered for the awards generally.

- Whilst the need to raise media interest more generally should be a primary aim, a number of interviewees felt exploring the potential to televise the awards would greatly help raise awareness of house design. Parallels were drawn with the Stirling Prize for architecture, which annually attracts television audiences of 1.2-1.4 million people. The Stirling Prize coverage on Channel 4 has successfully raised the profile of architecture, although it is recognised that this was achieved by making the award more controversial and exposing the judging process. The latter in particular has had resource implications however, as now many more schemes have to be visited. If an annual programme about the *Housing Design Awards* occurred, however, the interests of the wider public would be more fully met, and debates about the future of housing design would also be more firmly in the public domain.

In addition to these suggestions, a number of additional minor actions might also be considered. Firstly, the web site could be expanded to include all exhibition entries, and competition material could be circulated via a CD rom. A link could also be made between the travelling exhibition and regional debates and seminars organised by CABE and targeted at professionals and council members. In this way, the exhibition would provide a complementary tool to CABE initiatives related to widening awareness of contemporary approaches to housing design.

Moderate change

In addition to the changes suggested under a 'minimum change' approach, it is argued that the *Housing Design Awards* can undergo fairly moderate change yet strike an appropriate balance between retaining the existing ethos of the programme and seeking to counter some of its weaknesses.

- A key weakness is the belief that the awards are solely sponsored by RIBA. This could be overcome by stressing the particular roles of each of the four partners and how they collectively contribute to an award format which is not only oriented to policy and professionals but also reflective of wider industry and house consumer concerns. The role of the National House Building Council should be specifically enhanced, and their wider involvement sought, as it may be expected that this would make the award more attractive to developers. The need to balance the sponsor profile would perhaps also be helped if the exhibition was disengaged from the actual awards ceremony, as this acts to reinforce the view that the awards are entered by architects, judged by architects and the results disseminated to architects. If possible, the ceremony should also be hosted in a neutral location.
- A second weakness in the current Housing Design Awards format is the view that this is a rather elitist award dominated by professional interests and criteria. To make the award more balanced, a lay element could be introduced into the judging procedure and criteria. In relation to procedures, the experiences of the *Civic Trust*, *Evening Standard* or *Greenleaf Awards* offer models that might be considered. For example, the Civic Trust has a lay person on the judging panel of each scheme who will research public opinion locally and provide comments to the judging process.

- Criteria are currently fairly explicit for the *Housing Design Awards*, but they could be adapted to more fully reflect consumer concerns. Advice could be sought as to which criteria might be regarded as appropriate, although matters related to 'liveability', build quality, management and maintenance seem to be often quoted, and residents are most keen to comment on build quality as a dominant concern. In addition the award write-up and press releases for particular schemes could include comments from lay judges and, where possible, residents living in completed schemes.

- In exploring more effective dissemination, public relations professionals could be used to formulate a broader programme of opportunity to highlight the award winning schemes. Key messages emanating from the awards could draw out the issues of public interest, building on the human element and exploring, for example, what it is like to live in an award winning scheme or whether the scheme has helped people save energy and money. Such stories inevitably take the focus away from rather abstract discussions about housing design, but they may enable more effective targeting by PR professionals at relevant and more populist media, including the national press, colour supplements and lifestyle magazines. More 'people focused' content could also find its way into the current awards brochure.

Such changes would enable the breadth and inclusivity of the programme to be better appreciated: the *Housing Design Awards* are not simply an architecture award for architects. Exposing this assumption addresses a weakness, but does not reduce as a consequence the professionals' respect for the awards in terms of their quality of judgement and assessment. The potential of introducing a lay element into the judging process has more mixed implications. Whilst inclusion would help counter assertions that the *Housing Design Awards* are not geared towards industry and consumer concerns, there would be resource and time implications for any resulting judging procedure.

Fundamental change

As illustrated throughout this research, house builders want award programmes to have integrity, reflect consumer concerns in their judging, and be disseminated in a form that will be seen by their consumers. They will enter awards that they feel they have a chance of winning and they will target categories that match their output. An award that would reach out to mainstream developers would therefore result only from a more complete overhaul.

In addition to the above changes it is also suggested that a number of more significant changes could also be undertaken. Two specific approaches are proposed:

- the introduction of a wider range of specific award categories reflecting house builder product types;

- a regionally based benchmark award scheme.

Category Award: This would reflect approaches taken by the *What House? Awards* or the *National Home Builder Design Awards*. These have a wide range of categories for areas of design, such as for individual house types or particular development scenarios, which closely reflect developers' concerns. Aspects of the government's more specific urban renaissance agenda could be reflected in these categories or in a small number of awards given to reflect exemplary schemes.

Regional Benchmark Award: This approach is preferred and would reflect the *Civic Trust* or *Greenleaf Awards*. It would involve giving out a number of awards at the regional level for schemes that comply with a certain set of strict benchmark quality criteria, and with schemes going forward to a national award for more exemplary projects. This latter approach could retain the urban renaissance focus, but would award any scheme at the regional level that matched the quality criteria, including matters of density, mixed use, energy conservation, landscape and the role of the development in its context which would embrace sustainability issues. As schemes go forward to the national level the awards could reflect more fully the approach that is adopted now, drawing on the regionally recognised schemes to highlight particular exemplars.

The role of the regional press could also be enhanced. Volume builders appear to have a close relationship with the regional papers as they tend to reflect their markets and could be encouraged by links to promote their award winning status at the more local level. Companies could be informed where material about an award-winning scheme is to be reported upon, and they might be encouraged to include advertising to provide profile of other schemes in the region. The link between the regional and national dimensions of such an award would also allow the more general design agenda of the Government to be made relevant to the local level.

Realigning the awards through either of these methods may make them more attractive to private sector developers as they have more chance of winning. Equally, maintaining recognition of exemplary schemes as well as recognising achievement of particular standards means that a focus on quality and innovation is not necessarily diluted.

Of the two approaches suggested above, the former would be more resource efficient, but the competition would continue to be enacted at the national level. A regional benchmark award has more significant resource and time implications. The Civic Trust manage this by gaining the support of local planners and lay volunteers. The same approach could be considered for the *Housing Design Awards*, and it would bring planners, in particular, more fully into the process. The additional benefit of adding the regional award, however, would be that a wider range of regional schemes could be embraced, the awards programme would be able to develop a more bottom-up rather than top-down character. Housing developers would be more attracted to a scheme aligned more closely with its regional markets, with a greater commitment to localised forms of dissemination.

The Building for Life Standard

It is important to note similarities between recommendations arising from this research and the recent announcement in July 2003 of a new housing design Standard from the CABE/House Builders Federation/Civic Trust *Building for Life* initiative. The *Building for Life* Standard aims to reward builders of high quality new developments and increase awareness of the importance of good design and architecture to consumers buying new homes. Individual schemes are to be assessed against a number of criteria which very much chime with the Government's vision of good place making as outlined in the *Sustainable Communities – building for the future* (ODPM, 2003) including design and construction, environmental impact, public space and creating a sense of community.

The standard is awarded to house builders and housing associations (NB not specifically architect practices) and has two principal categories: silver and gold. A silver scheme will be awarded to exemplary schemes and will need to fulfil 70% of the criteria. A gold scheme will need to meet 80% of the criteria and must demonstrate the highest standards of architecture and all round sustainable development. Schemes that meet 60% of the criteria are considered for the *Building for Life* library of best practice. An annual platinum award may also be given if an exemplar scheme fulfils over 90% of the criteria.

The arrival of the *Building for Life* Standard therefore introduces an additional important factor in considering the potential future role and function of the *Housing Design Awards*. In many regards, many of the possible changes recommended – and certainly the more substantive ones – would in effect move the *Housing Design Awards* towards the aims and scope of the *Building for Life Standard*. It may even be the case that in using the *Building For Life Standard* to reward exemplar design clearly in line with Government policy, the longer term purpose of the Housing Design Awards becomes less clear.

Final comments

It is important to reflect in final conclusions that traditionally design awards have not had a significant or direct influence on the form of residential development, and as such the mechanism – however formulated – offers only limited opportunities for Government to promote and encourage better housing design. The *Housing Design Awards* in particular, whilst being a very successful awards programme, do not reach out as might be hoped to many private sector developers, with most interest for entries coming from architects working for a whole range of clients.

The relative importance of awards in encouraging and promoting better design aside, the research has highlighted a shared view amongst all those involved in the house development industry that there has been a significant improvement in recent years in the quality of some types of housing development. This is particularly the case with new types of product emerging to reflect government planning policy agendas including city centre regeneration and brownfield land redevelopment. There is also a clear driver from the consumer side, with improving design awareness amongst the house buying public. This seems to represent a certain synergy between housing design practices, the urban renaissance agenda alongside cultural trends and interest in new ways of living. Such trends are encouraging all developers to commit some area of their practice to a renewed interest in the potential of design. In particular the need to use difficult brownfield sites and develop them to a higher density, necessitates greater use of architects in order to maximise the value of sites. There is also evidence to suggest that even a number of the volume developers have altered their approaches to development to allow greater differentiation within their standard house types so that they can contribute to creating better housing layouts. Such trends are restricted to a small number of schemes typically shaped by the specific circumstances of their context and where the market will support such investment in design.

The *Housing Design Awards* reflect rather then drive these trends, and provide a professional, well-regarded assessment of emerging forms of innovative housing design. In particular, the awards highlight a level of commitment to innovation in the form and style of housing, although such innovations do not fully extend towards greater awareness of forms of housing that might encourage more sustainable lifestyles. This research has shown that other awards have their own particular strengths and there could be merit in considering possible changes whether through mechanisms to involve private developers more fully, reflect lay concerns a little more, or target dissemination activity at a wider audience.

Such choices about the future of this award programme can only be made with a firm view about the values underpinning it. One important value must be to prove the suitability of Government housing design policy, and this focus should not be lost. Current development practice seems also, however, to suggest that in the future schemes will continue to emerge that such an award programme can very usefully reward and bring to the attention of us all.

Appendices

Appendix 1

Awards for Housing Design in England

Appendix 2

Other Housing Related Awards Running in the UK

Appendix 3

Interviewees

Appendix 4

Award Winning Schemes Used as Case Studies

Appendix 5

Bibliography

Appendix 1

Awards for Housing Design in England

National Home Builder Design Awards

Sponsors: The Mail on Sunday and Standard Life Bank (formerly the Britannia Building Society and The Daily Telegraph). Also mentioned are the RIBA and Smart New Homes web site. Individual awards have been sponsored by Ideal Standard, Smeg, Corgo, BAC, Techaus, Gaz Co, Vent Axia, Marshalls, Porcelanosa and Nolte. The award always seeks sponsorship from a national newspaper.

History: The awards were established in 1981. Eligibility for inclusion was changed in 1999 to only schemes designed by registered architects working for private house builders producing housing for sale. The restructuring occurred under the influence of the Chairman in response to views that the criteria for eligibility were previously unclear, the process regarded as insufficiently selective, and the Awards simply chose the 'best out of what was offered'.

10 new categories were introduced through the restructuring, and since 1999 every entry must be signed by a named architect on the register of the Architects Registration Board in the UK or an equivalent body outside the UK. Such an architect may be working as a consultant to, or on the staff of, a house builder on the register of the National House Building Council or the House Builders Federation, or an overseas subsidiary of such a builder.

Focus of the Awards: Schemes designed by registered architects working for private house builders producing housing for sale.

Submission: Submissions are expected to include: indication of category (see below), details of home builder, architect, landscape architect/designer, description of the scheme and a sales brochure, location plan and site layout plan (A4 if possible), floor plans of the buildings, photographs, entry fee (£235).

Judges (2002): 3 architects, a representative from the HBF, a developer, 3 architectural journalists, a representative from Design for Homes, a consultant and a presenter.

Procedure for Judging: The editor of *Planahome* appoints the judges. Since 1999, judges have numbered between 6-8. A balance is sought between architects and intelligent lay people, so that the jurors are not simply 'architects judging architects'. Typically, the 'lay' jurors express their preferences first.

Companies submit entries (c. 200-300), and the judging panel carries out a preliminary 'cull', deleting those not up to scratch. Each member of the jury looks at every submission and makes a comment – 'a very fast process'. A collective view is reached on a second round, which are then rated to provide (as a third round) a short list, on the basis of which the final decisions in relation to each category are decided.

Most schemes are not visited, except the short-listed schemes for the Manser Medal (see below).

Awards Given: Categories include:

- Best house in England
- Best house in Scotland
- Best house in Wales
- Best house in Ireland (none of previous categories are one-off houses)
- Best Apartment Building (new build)
- Best restoration and conversion (to housing or from house to flats)
- Best use of a brownfield site (new build)
- Best house of the future
- Best retirement development
- Best partnership housing development
- Best landscaping of a development
- Best housing development of the year (at least 10 dwellings)
- Best housing project of the year (at least 10 dwellings due to start or be completed during the year)
- Mail on Sunday readers award (see below)
- Manser Medal for the Best One-off House in the British Isles.

Criteria: Criteria generally relate to the creation of buildings and layouts which provide a good, practical living environment, with space efficiency, and adequate light. Estates should be more interesting than those resulting from the application of by-laws. Layouts at higher density should preserve privacy.

Individual categories also have specific criteria:

- Brownfield schemes are judged on environmental quality before and after development.
- Best house of the future is judged on innovative design and sustainability in terms of construction and materials, conservation of energy and water, adaptability, accessibility and the use of technology.
- Best retirement development is judged on external and interior design, layout, landscaping, security and the provision of services and communal facilities.

- Best landscaping of a development is judged on the quality of the communal spaces and of private gardens of properties.
- Best housing development of the year (+10 dwellings) is judged on overall appearance, layout and landscaping, and the external and interior layout and design of the dwellings.
- In 2003 the Mail on Sunday sponsored a reader category where readers vote for one of 10 houses chosen by the judges.

Awards and Dissemination: The results are disseminated via:

- Judges Report
- *The Mail on Sunday*
- Annual *Planahome* publication/magazine available in the high street
- Annual Best of *British Homes* publication
- The awards pages of the RIBA website
- Web site www.nationalhomebuilder.com

In 2002 the Manser Medal winner, commended entries and other entries of merit were also featured in *The Sunday Times* and the *Architect's Journal.*

Civic Trust Awards

Contact: The Awards Office, The Civic Trust, 17 Carlton House Terrace, London, SW1Y 5AW. Telephone: 020 7930 0914, awards@civictrust.org.uk

Sponsors: Special awards (see below) are sponsored by Waitrose, the Civic Trust, Woodhouse UK plc, English Heritage, ODPM, English Partnerships, Thorn Lighting and OSRAM, Welsh Development Agency, and the Countryside Agency

History: Since being established in 1959, the Civic Trust has presented 1082 awards. The awards were established only a few years after the foundation of the Trust in order to promote examples of good development practice. They were introduced within a context of quite rapid changes to many towns and cities, and the Trust was keen to provide a mechanism for both promoting and encouraging good design within such a context.

Organisation: The Civic Trust has an awards office. Awards are given on a two-year cycle: one year for projects in major city areas and one year for schemes in the countryside and country towns. The awards include up to 400 volunteers a year involved in judging. Operation of the awards costs approximately £180,000.

Focus of the Awards: The awards are given to 'outstanding examples of architecture and environmental design and are unique in that they take into account the benefit each project brings to its local area as well as considering the quality of its design.' In particular, awards are given for outstanding projects in the fields of architecture, planning, environmental design, landscape, public art and urban design.

Submission: Submissions include: entry form, short description of the project, a location plan, a plan of the project, at least 5 photographs and an entry fee.

Judges (2002): The National Panel comprises: 4 architects, 2 planners, a landscape designer, an engineer. Local judging panels are locally lead by an architect, and each includes a planner, a lay person and a disabled access expert.

Procedure for Judging: The judging procedure takes 5 months.

Stage one (October)
The entries are organised with the help of participating local authorities who collect the submissions in their local area. Local Authorities can vary in their commitment. The good Authorities tend to have good key people or a generally good design culture (e.g. Manchester and Glasgow).

A member of the LA arranges for a judging team to visit the entries in each area. The visit should not include anyone who was involved directly in the planning/designing of the project. The team of assessors for each area is nominated by the Civic Trust. Each team is lead by an architect with no links to the area. The team also includes a local planner from the area, a lay adviser who lives in the area, and where possible an expert in disabled access. The lay advisor seeks the views of locals on the scheme and reviews the local press. This group reports back to the Civic Trust. The team leader writes the report and has the final say about the award, although this should be as a result of a group process and significant disagreement should be highlighted.

If a scheme in the area has not been submitted for an award, depending on agreement with the owner, it can also be recommended by the panel, visited and included.

Stage two
This second stage involves a national panel of judges reviewing the recommendations from the area teams. The panel includes experts in the fields of architecture, planning, landscape, conservation, access and sustainability and meets to ensure comparability of judging from across the UK. The panel has the ability to overturn an assessor's recommendation or may wish to discuss a particular scheme with the assessor. The group also agrees which special awards will be given unless specialist judging is required.

Awards Given:
The titles of the special awards seem to change; previous special awards have included:

- *Centre Vision Award*: given to a scheme that contributes most to the improvement of a town or city centre.
- *Sustainability Award*: given to the scheme that demonstrates the best application of the principles of sustainable design and construction.
- *Landscape Award*: best landscape treatment of a site or area.
- *Access Award*: given to a scheme that best reconciles the access needs of people with disabilities with the conservation requirements of a historic site or building.

- *Urban Design Award*: Given to schemes that make the best contribution to the local environment as a whole and includes more specific criteria. This award was only open to English schemes in 2002.
- *Partnership Award*: Best regeneration or development involving partnership between private, public and community sectors.
- *Exterior Lighting Award*: given to the best exterior lighting scheme that most successfully respects and enhances the architectural fabric of a building or built environment.
- *Rural Design Awards*: for schemes that are sustainable in terms of their location, construction and long term use, that benefit the community and respect the character and distinctiveness of the locality. Awards are given to Market Towns (2,000-30,000), Vital Villages (<10,000) and agricultural buildings. Only available to schemes in England.
- *Welsh Regeneration Award*: for the regeneration project in Wales that achieves the highest quality of design and promotes broader sustainability objectives.
- Award for outstanding contribution to the quality and appearance of the environment. The scheme should be truly outstanding in every respect and demonstrate the highest quality in terms of architectural/design intent and execution.

Commendations are given to schemes that make a significant contribution to the quality and appearance of the environment, and mentions are given to projects that bring a notable social, cultural or economic benefit to their communities and where the architectural/design quality is above average.

Criteria: Civic Trust includes a detailed statement of criteria for judging. Schemes must be complete (including mature landscaping). Recognising the importance of context is key, and well designed schemes that do not enhance their context will be rejected. Schemes must be visible to the public all year round and disabled access needs must be considered. The background of schemes should be researched, in particular to determine what existed on the site before the scheme. Judges should also determine whether a scheme is likely to weather or suffer poor maintenance.

Awards and Dissemination: Winners of the 'special awards' and 'awards' are given a certificate and a plaque. 'Commendations' and 'mentions' receive a certificate. *Building* magazine publishes a special supplement outlining the award winners every year. The awards office/Civic Trust Press Officer sends out press releases to national and local media, and to award winning local authorities for their own use. The awards target and successfully receive local coverage.

Evening Standard New Homes Awards

Contact: Martin Smith, Residential Property, Classified Advertising Department, Evening Standard, Northcliffe House, 2 Derry Street, Kensington, London W8 5EF.

Tel: 020 7341 8232

Email: martin.smith@standard.co.uk

Sponsors: Evening Standard. (The Evening Standard, is an important advertising vehicle for house builders in London)

History: The awards were established in 1991.

Focus of the Awards: New homes built within the readership area of the Evening Standard (the South East of England).

Organisation: Hall Associates organises and runs the Award on behalf of the Evening Standard, and were responsible for developing the judging criteria. Each entry pays a fee of £150 – the remainder of the funding is provided by the Evening Standard (the precise sum is confidential).

Submission: Entry is open to all house builders active within the readership catchment area of the Evening Standard (essentially London and the South East). All submission material must be in an A4 binder and contain the following:

- Brief description of the property or development.
- Detailed area and site location maps.
- Floor plans and drawings.
- Consumer specifications.
- Photographs of the interior and exterior.
- Overall site photographs (development categories).
- Short description of the company.

Judges: Initially there are eight judges including an architect (Piers Gough) and a number of feature writers. Subsequently the paper invites readers to become judges – 26 are selected for this purpose. Readers must not have vested interests, and are vetted accordingly.

Procedure for Judging: The team of judges (approximately 8) assesses all the entries and establishes a short list in each category. Short listed properties are visited, and a further short list of 10 schemes in each category is produced. Evening Standard readers active in the markets covered by the categories are invited to form part of the Judging Panel. Under an independent Chairman the judges for each category choose a 'winner' and a 'highly commended' amongst the schemes. Normally each reader judge will be involved in a couple of categories. These judging panels make the final decision but they do not visit schemes.

Given this expert/lay mix within panels, the awards seek to represent the views of 'the buying public' on the basis of shortlist put in place by professionals.

Awards Given:
Award categories are:

- Best luxury new home.
- Best new family home with 5 bedrooms.
- Best new family home with 4 bedrooms.
- Best new family home with 3 bedrooms.
- Best new conversion.
- Best new apartment.
- Homes and Property London Lifestyle Award.
- Best new development by a housing association.
- Best new development.

Two sets of awards are given in each category: house builders who construct over 100 units a years and house builders who build under 100.

The London Lifestyle Award was introduced in 2001 (for the best design led property or development within the Greater London area that makes the most innovative use of space, energy conservation, materials and technology).

Criteria: Judges are given advice about criteria to be used in selecting schemes. Those listed on the entry form include:

- Quality of design.
- Utilisation of space.
- Location.
- Environmental Factors.
- Fitness for purpose.
- Quality of finish.
- Value for money.
- Use of materials.
- Energy efficiency.
- Innovation.

These criteria are interpreted in a flexible manner and adapted according to house price.

Dissemination: Prizes include a trophy, wooden plaques and flags. Winning and highly commended schemes in each category are entitled to fly the award flag. Awards are given out at an exclusive lunch. Of course, the principle dissemination route is through the Evening Standard Homes and Property section!

Greenleaf Awards

Contact: New Homes Marketing Board, 56 – 64 Leonard Street, London, EC2A 4JX

Sponsors: Design for Homes, *Daily Mail*, Hanson, Wildfowl and Wetlands Trust, Civic Trust, Tree Council, English Partnerships, National Urban Forestry Unit

History: Established in 1986 by the New Homes Marketing Board

Focus of the Awards: The *Greenleaf Awards* focus on the spaces between the buildings, recognising best practice in the context of new housing landscaping/landscape design. By doing this the awards seek to:

- help raise standards of landscape design throughout the industry;
- raise house buyers expectations by demonstrating the importance of landscape;
- and help to raise the reputation of the house building industry with planners, policy makers and politicians.

Organisation: The awards are run annually and scheme submissions are required by September each year. The awards cost about £35,000 – 40,000 annually to administer. Funding principally comes from the House Builders' Federation, although other sponsors include Hanson, and there is sponsorship in kind from the *Daily Mail.*

Schemes must be at least 3 years old. This is because:

- schemes have had a chance to mature and planting can become established;
- the relationship between private gardens and public landscapes will have settled down;
- there will be evidence of poor management and maintenance;
- damage to vulnerable elements (water courses and mature trees) during the construction process will be exposed;
- issues such as litter, dead trees etc will be more apparent.

Submission: Submissions are initially by application form. The form requests information about:

- Name and address of the development.
- Date that work started and was completed.
- Type of landscape prior to development, including established landscape and barren landscape.

- Location: urban, rural, greenfield, brownfield, other.
- Size of the development (homes): 5-10, 10-30, 30-100, 100+.
- Professionals involved: architects, landscape architects, ecologists, tree surveyors, engineers.
- Checklist of features influencing the design: mature trees, streams and ponds, fine views, site toxicity, noise, hedgerows, historical structures, unattractive views, redundant buildings.
- At least one recent photograph with description.
- Who manages the landscape: in-house, original contractor, landscape contractor, LA, residents.
- Contact details.

Judges: It is not a technical award and judging is not professionally led. The landscape is being judged for liveability – whether it feels good and looks well cared for. A small committee made up of 2 people from the House Builders Federation, 2 from the *Daily Mail*, the editor of *Housebuilder Magazine* and the Chairman undertakes initial judging. Additional judges are derived from the sponsoring organisation.

Procedure for Judging:

Stage One
Schemes are initially judged on paper. Schemes not conforming to the 3 year rule and which do not show that care and consideration has been given to the landscape are set aside.

Stage Two
For proceeding schemes a more detailed application form is sent out, to be returned by 31st January, with additional photographs.

Lay people from the range of sponsors visit every site. For example, a sponsor's sales team will be trained up and they would go out with a tick box form and visit allocated developments. 2 people visit a scheme. The issues are quite straightforward – has the grass been cut, is it attractive, are there big dead trees in the scheme? The main judges then just see the comments submitted on the 2 sheets and compare these with the photos. If comments contradict, they might follow up with questions to the visitors. For the specialist awards someone will then visit the site with more knowledge.

A ceremony occurs in May at the Daily Mail head offices.

Awards Given:

- A *Greenleaf Award* given to a number of schemes that conform to a standard.
- Hanson sponsored overall award winner.
- National Urban Forestry Unit Award: for planting beyond the development, and to encourage strategic greening.
- Civic Trust Award: must link closely to its surroundings, make a sense of place where residents can readily become part of the local community.

- Wildlife and Wetlands Trust Award: must show how water can be managed as an asset in new development.
- Tree Council Award: for protection and integration of mature trees.
- English Partnerships Award: for reusing previously developed land and for overall environmental improvement.
- Design for Homes Award: for sustainable landscaping.

Criteria for Judging:
"Liveability". (From a judging form) Is there any litter in the shrub beds yes or no? Is the grass uncut? Are there trees with dead branches in them?

"Essentially it's about whether the landscape is well cared for and growing after 3 years."

Awards and Dissemination:

- Award winners get a banner and certificate.
- The *Daily Mail* promotes the award through its regional newspapers.
- They link up the award winners with the regional/local newspapers in their local areas.

Housing Design Awards

Contact: The Awards Secretary, Housing Design Awards Office, C/O Birmingham and Midland Institute, Margaret Street, Birmingham, B3 3SP

Tel. 0121 233 2531

Sponsors: Office of the Deputy Prime Minister, National House Building Council, Royal Institute of British Architects, Royal Town Planning Institute, (also *Housebuilder Magazine*, Building for Life Initiative (supported by the House Builders Federation, Commission for Architecture and the Built Environment and the Civic Trust))

History: (taken from the Awards web site)

In 1947 Aneurin Bevan, then Minister for Health with responsibility for housing, announced that his Ministry would be giving annual awards for public housing design and layout. After consultation with the RIBA, awards committees were set up for each of the then four English regions, presenting one medal each for an urban and a rural scheme, except for London where the medals were for one new scheme, and one reconstruction.

The initial scheme ran until 1955, and during this period awards reflected the increasing importance of the urban reconstruction and New Towns programmes, including the first high rise developments. In 1960, the scheme was reconstituted as the Ministry of Housing and Local Government (later DoE) Good Design in Housing Awards, sponsored jointly with the RIBA. The new scheme covered both private and public sector housing, reflecting the emergence of imaginative speculative developments. The next ten years showed a significant shift of emphasis away from public sector, high density, high rise and private sector low density, low rise to compact housing schemes in both sectors. In 1981 the NHBC joined the DoE and the RIBA as sponsors, to create the *Housing Design Awards*. The Awards became a biennial event and the public and private sector categories were abolished, reflecting their increasing convergence.

The last ten years have seen a further shift towards diversity in housing, away from the rigid categories employed in earlier years. Mainstream housing today needs to cater for a very broad range of needs. The new streamlined structure of the Awards is designed to accommodate this diversity in as flexible a manner as possible.

Focus of the Awards:

- Schemes must have four or more dwellings.
- Included is housing design from the private or public sector and including new build, conversion and renovation in England and Northern Ireland.
- Schemes including a non-residential element may be entered so long as housing constitutes the major element. Nursing homes are not eligible.

Organisation: The current annual cost of the Awards is approximately £105,000. Of that, just over £25,000 goes on the printing cost of the Winners Book and placing an electronic version on the website, which is paid by ODPM. The proposed income for 2003 from sponsorship of three Special Awards and of the Exhibition is £26,000. Some £13,000 income comes from entry fees and sales of the Winners Book. The promoters meet the remaining costs.

Submission: Submissions should be presented on one side of an A1 board (landscape format). They should include:

- a location plan showing the site and surrounding area;
- layout plan with north point and scale, showing the relationship between the proposed building to any others;
- typical dwelling plans and elevations;
- in conversion and improvement, photographs of the existing building, and in other cases photographs of the site and/or adjacent developments;
- text setting out the brief, town planning context, materials used, landscape treatment, timetable and special features;
- large schemes (+50 houses) can submit two boards for a higher fee;
- a copy of the planning permission and planning officers report should be submitted.

Judges: 2 representatives from each sponsoring organisation: NHBC, ODPM, RIBA, RTPI. Observers from the administration side of the promoters may also be invited.

Procedure for Judging: There are two stages:

Stage One
Stage One takes place over two days in March and involves the selection of projects and completed schemes for exhibition. There are usually about 200 submissions.

Each sponsor contributes two assessors. These assessors are split into 4 teams containing different disciplines which each sift a quarter of the entries. Sifting is to pull out clear "non-starters." Displayed boards are assessed using the criteria and if they are not good enough they are turned face down, but if there is disagreement the board is turned face up again. After this is done each team inspects the results of others and has the power to turn boards over as necessary. The process is repeated until an exhibition total is reached (70 boards). Schemes selected for exhibition should demonstrate range of type.

From the projects chosen for exhibition, 8-10 project award schemes and 14-16 competed schemes are chosen to be visited over 2 "2-day" visits. Assessors also then select candidates for the special awards

Stage Two
Each sponsor contributes one assessor in addition to the Committee Chair. The group visits the short listed awards and determines the final 3/6 winners. Judges aim to spend an average of three-quarters of an hour at each scheme, speak to residents, and ask to be given access to at least two representative dwellings. Each assessor decides which schemes to award and they can also identify 5-6 schemes "Worth a Detour." The final decisions are combined in a matrix and some consensus hopefully then emerges. Once awards are fixed, special awards are also determined.

Special Awards
Representatives of the sponsors of the awards determine these. They are given information by the above stage two assessors and usually the assessors will also make a strong recommendation.

Awards Given:

- *Project Awards*: projects that have detailed planning permission, but are not yet complete
- *Complete Schemes*: Completed schemes between specific dates. Previous project awards are automatically short listed.

Special awards:

- *Building for Life Award*: for the best examples of a private sector development at over 70 dwellings per hectare.
- *House Builder Award*: For the best example of a private sector development at between 30 and 70 dwellings per hectare.
- *Housing Corporation Award*: For the best example of affordable housing

Criteria:

- Relationship to surroundings and neighbourhood.
- Response to site constraints and opportunities.
- Layout grouping and landscaping.
- Planning of roads and footpaths.
- Handling of garages and car parking.
- Attention to safety, security and accessibility.
- External appearance and external planning.
- Sustainability in construction, planning and layout.
- Finishes, detailing and workmanship.

Awards and Dissemination:
Dissemination is via:

- Award winners get a plaque and/or a certificate.
- An exhibition of submission boards in the RIBA, which this year will subsequently go on a national tour.
- An awards ceremony.
- An annual "Winners Book" publication/catalogue.
- *Pick of the Crop* publication.
- Web site.
- A number of schemes were also put on *The Guardian* web site.
- Annual article for *Housebuilder Magazine*.
- The awards are also covered by the national and professional press.
- The exhibition might be circulated more widely by CD Rom.

What House? Awards

Contact: Blendon Communications Ltd, 207 Providence Square, Mill Street, London SE1 2EW, email Jo Walsh: jw@blendoncom.com

Sponsors: *What House?* magazine, *The Sunday Times/Daily Telegraph*, De Dietrich

History: *What House?* Awards were established in 1981.

Focus of the Awards: Houses and schemes built by private sector developers.

Organisation: The awards occur annually. House builders can submit any number of entries.

Submission: Submission requires a form which must contain details of the following:

- Category for which the scheme is submitted (see below).
- Name of the scheme.
- Postal address of the development.
- Date of completion of building work.
- Name of the architects (where applicable).
- Contact name.
- Market prices should be included with all entries.
- 2 slides of the development for use in the Gala Presentation lunch.
- Internal and external, site layout, location and landscaping plans (all at A4 or A3 folded).
- Photographs.
- A maximum of 4 pages outlining the details of the entry.
- Details of room sizes and total floor areas.

Judges: Architects – 5; Journalists – 6; Landscape Architect – 1; Surveyors – 1.

Procedure for Judging: The judges produce a short list of the best entries in each category. Each short listed scheme in each category is then visited and a report is written before the judges make their final decision. Where judges cannot make a decision, entrants may share a prize.

Awards Given:
All winners receive a plaque. Gold, silver and bronze awards are given in each category

Award Categories:

- Housebuilder of the Year Best Volume House Builder (+1000 units).
- Best Medium Sized House Builder (100 – 999 units).
- Best Small House Builder (less than 100 units).
- Best Development; Best Luxury Development.
- Best Starter Home; Best House; Best Luxury House.
- Best Apartment.
- Best Renovation.
- Best Interior Layout.
- Best Future Home.
- Best Retirement Development.
- Best Brownfield Development.

- Best Partnership Development.
- Best Exterior Design.
- Best Landscape Design.
- Best Energy Saving Development (sponsored by English Partnerships).

Criteria: An array of criteria is referred to in the documentation with each category having a descriptor, some examples include:

- *Best Development*: "stylish mix of original design, top quality site layout and landscaping, a variety of external elevations and plenty of design flair inside and out."
- *Best House*: "Best House is a middle market house with that something extra – the high quality family home – a standard house type of exceptional standard... What judges are looking for is a home that combines comfortable family living at a realistic price, but with the edge to lift it above the competitors..."
- *Best Renovation*: "enterprising development of dated or derelict property. Exterior or interior appearance, as well as interesting adaptation of original features will be considered."
- *Best Landscape Design*: "Judges are looking for a sensitive, yet practical, response to the site conditions, ensuring that the overlook of the scheme is easy on the eye with landscape features to match the quality of the homes. Entries should be developments that make maximum use of natural features or create a brand new landscape to complement or enhance the street scene."

Dissemination:

- *What House?* publish an annual supplement.
- Covered in *Show House* – trade magazine for the sales and marketing arms of house building companies.
- National newspaper supplement.

Appendix 2
Other Housing Related Awards Running in the UK

British Housebuilder Awards

Awards are given for *quality* of service to customers. The award is based on a market research format with awards being given to companies performing well in certain areas of service provision. House builders register with the competition and pay a fee. For this fee, the company ultimately receives a computer print out of their company's performance. This is based on responses to questions about the company from house buyers. Results are compared to national averages for a range of questions (see below). The printout also includes the company's ratings against the top five companies in the respective size band. Awards are given in a range of categories related to size of the house building company. Questions relate to:

- Accuracy of all literature.
- Impressions on first visiting the development.
- Helpfulness of sales staff.
- Helpfulness of all other staff on the development.
- Builders advice concerning energy efficiency aspects of construction.
- Help and advice received during the home buying process.
- Meeting building deadlines.
- Condition of the home on moving day.
- Condition of paths, drives and garden on moving day.
- Efficiency of the builder in rectifying faults after occupation.
- Efficiency of builder in rectifying faults to date.

There is an awards presentation and Express Newspapers run a number of special editions on the awards. In addition a press pack is circulated to regional press outlining award winners and including details of new developments.

Building Homes Quality Award

The award, first given in 2001, is for the business performance of the house building industry. No design factors are taken into account. Awards given include: Best training/people development strategy; Best customer satisfaction improvement strategy; Best quality improvement strategy; Best change strategy; Best options and choices initiative; Best health and safety approach; Best company wide sustainability strategy; Best approach to partnering/ supply chain management; Best approach to partnering with sub-contractors; Best building efficiency initiative; Private sector house builder of the year; Affordable sector housing provider of the year.

BURA Award for best practice in regeneration

This award, sponsored by the British Urban Regeneration Association and English Partnerships, rewards urban regeneration initiatives. 6-7 awards are given annually. Housing can feature in awards as the focus for a project or more likely as a part of a larger scheme/strategy, although housing design is typically not the focus. Award winning projects should:

- Make a successful economic contribution to the regeneration of an area and be financially viable.
- Act as a catalyst to further regeneration.
- Contribute to community spirit and cohesion by raising levels of confidence in the long term living and working environment.
- Contribute to building the capacity of local people.
- Contribute to environmental sustainability; be completed to the point where there is a track record of success.

BURA Award for Urban Renaissance

The award was presented for the first time in 2002. The award is for urban regeneration initiatives of all types and sizes in England that can point to demonstrable results. The award will have a different theme each year, although the theme will be derived from the urban renaissance agenda. In 2003 projects must contribute to making towns and cities better places to live and work. Ways in which a project could demonstrate this include:

- Involvement of the entire community
- Creation of safer, cleaner and better managed streets
- Implementation of a local approach to transport and traffic management
- Creation of a new public space, which is well designed, easy for all sectors of the community to get to, and in which no one feels excluded.

One of the award criteria is high quality design, and the award could relate to a residential environment. The award is sponsored by the Urban Policy Team at ODPM.

BURA Charitable Trust Community Regeneration Awards

The award is sponsored by English Partnerships and has been established for 8 years. The awards aim to promote outstanding examples of community based regeneration. In particular they aim to: ensure the involvement of local people as equal partners in development and implementation of regeneration programmes; maximise the positive impact of regeneration schemes on peoples' lives; and promote the exchange of successful ideas and formulate principles of best practice. Projects must:

- Be inspired by the community.
- Contribute to the raising of community spirit and improve the quality of life of local people.
- Be viable and sustainable in the long term.
- Be accountable and responsive to local stakeholders.
- Address local needs identified by the local community.
- Demonstrate a proven track record of success.

Europan

Europan is sponsored in the UK by the ODPM (English Partnerships, RIBA, and the Architecture Foundation are also referred to in the documentation). This is an annual pan-European architectural design competition. The competition aims to encourage professional European architects under 40 to engage in the design of housing in European cities. The UK became a full member of Europan 4 years ago, and subsequently offered competition sites for consideration. In Europan 6 there were 67 sites in 23 European countries. Each winner will be commissioned to further develop the competition site. The specific aims of the competition are to:

- Increase depth of knowledge of, and reflection on, both housing issues and urban planning.
- Help young architects to develop and promote their design ideas.
- Help towns and developers find innovative architectural solutions to the problems that they face.
- Promote the building of experimental projects.

Inside Housing UK Housing Awards

The awards have been running for 6 years and were launched to celebrate excellence in the provision of social/affordable housing in the UK. Awards are given for: the best funding solution; the best regeneration scheme; the best campaign to promote social housing; and also good practice in: creating sustainable communities; resident involvement; housing management/maintenance; multi-agency working; promoting diversity; and recruiting and developing housing professionals.

Each award area has a specific set of criteria used in judging. For example, the award for creating sustainable communities is judged as follows: evidence of opportunities for community empowerment; multi-agency working and partnership; demonstrable difference in health, training and employment opportunities, crime and anti-social behaviour; reduced void turnover and increasing demand to live in an area.

National Home Improvement Council Award

The awards are given to modernisation and renovation schemes. The aim is to draw the attention of central and local government, social housing providers, homeowners, landlords and the media to the ongoing importance of improvement and renovations to housing. The awards have been running for 25 years. Awards are given for the following categories:

- Completed modernisations by local authorities.
- Schemes that help private owners improve their own properties.
- Housing Association schemes.
- Innovative training in roofing.
- Best use and most interesting development of a brownfield site or recycled building by either a public or private sector organisation or developer.
- Award for innovation in the management of gas safety.
- Innovation in bathroom design.
- Excellence in a roofing project.
- Projects which have made a significant contribution to the elimination of fuel poverty.

Each category has its own criteria. For example, the criteria for the award for the reuse of a brownfield site states: "The judges will be looking for innovative projects which particularly utilise existing buildings, homes and adjoining land to provide good quality housing. The scheme should also help the adjoining community and be able to be replicated elsewhere."

RTPI Planning Achievement Awards

The awards were established in 1978 by the Royal Town Planning Institute, and are given for '*outstanding group achievement in town and country planning*' in Britain.

The awards aim to reflect the diversity of planning work, and can include schemes that are urban or rural, large or small in scale, new developments or regeneration schemes and whether promoted by the public, private or community sectors. The awards look beyond purely physical outcomes to recognise innovative plans, planning processes, techniques, programmes of public involvement, information systems and public service improvements. Housing developments have formed part of award winning schemes in previous years.

Saltire Society Housing Design Awards

The awards were initiated in 1937, and are Scotland's main awards for housing design. The awards are sponsored by the Saltire Society, The Scottish Executive, Communities Scotland, National Housing Building Council (Scotland) and Scottish Homes.

The awards focus on new housing schemes in Scotland. There is a particular interest in receiving schemes from the private sector as a result of the concern expressed in the Architectural Policy for Scotland over the quality of private housing developments. They aim to encourage projects that address issues of sustainability, and encourage private sector projects that form part of regeneration areas. Awards are given for new housing, new individual houses, and housing restoration, reconstruction and conversion.

Criteria for judging the new housing category are: layout design; grouping of dwellings (5 or more) and their external treatment; relationship to the surrounding landscape; the use of colour and materials; general trim, degree of care and maintenance by occupants or the commissioning body; and the interior planning of houses relative to their type and location.

Welsh Housing Design Awards

The awards are sponsored by a wide range of governmental and professional organisations including: the National Assembly for Wales; the Design Commission for Wales; the Chartered Institute of Housing in Wales; RSAW Housing Design Group; House Builders' Federation; Royal Town Planning Institute; Welsh Local Government Association; and Civic Trust for Wales.

This is a new award, initiated in 2002, and will be given every 3 years (4 awards were given in 2003). The awards will be given to individual houses, large and small developments, and nurse and student accommodation. Mixed-use schemes are eligible, provided that housing constitutes the main element.

Criteria for judging include: architectural ambition, ideas and innovation; sustainable qualities of the design; accessibility; security; fitness for purpose and user satisfaction; standard of construction; context within the landscape or urban scene; economic impact and value to the community; and marketability.

Appendix 3
Interviewees

Award Organisers

Professor Chris Bains, Chair of the *Greenleaf* Awards
Mr Rupert Bates, Editorial Director, What House? Magazine
Mr Michael Hall, Managing Director, Hallmark Associates
Michael Manser, The Manser Practice, London
Mr. Stephen Mullin, Rapporteur for the *Housing Design Awards*
Graham Pye, Chairman, National House Building Council

House Builders

John Andrews, Land Director, Crosby Homes (Special Projects) Ltd.
Neil Armstrong, Marketing, Copthorn Homes
Louise Beesley, Sales and Marketing, Taywood Homes
Mr A Brown, Managing Director, Cala Homes
Fran Connop, Marketing Director, Crosby Homes (Special Projects) Ltd.
Mr A Dowse, Principal, Environ Country Homes
Trisha Gupta, Countryside Properties
Jason Honeyman, Managing Director, Barratt Eastern
Andrew Hurst, Marketing Manager, Taylor Woodrow
Nick Johnson, Executive Director of Development (Manchester), Urban Splash
Mr G. Jones, Land Director, Cala Homes
Tom Jones, Chief Architect, Taylor Woodrow
Lorraine Kendrick, Sales and Marketing, Crest Nicholson Residential (SW) Ltd..
Ms S Parry, Director of Sales & Marketing, Cala Homes
Ms P Reardon, Director of Sales & Marketing, Sunley Estates
Mr D Smith, Director of Sales & Marketing, Octagon Developments Ltd.
Paul Talbot, Design Executive, Crest Nicholson Residential (SW) Ltd.
Mr G. Taylor, Development Director, Linden Homes (Chiltern) Ltd.
Ms H. Todd, Director of Sales & Marketing, Linden Homes (Chiltern) Ltd.
Mr C. Tutt, Chief Executive, Octagon Developments Ltd.
Stephen Usher, Managing Director, Crest Nicholson Residential (SW) Ltd.
Nicola Wallis, Residential Sales Manager, Urban Splash
Mr A Wibling, Chief Executive, Sunley Estates

Architects

Mr M. Adams, Director, John Thompson & Partners (Architects)
Mr S Beck, Partner, Clague (Architects)
Teresa Borsuk, Project Architect, Pollard Thomas and Edwards Architects
Keith Bradley, Senior Partner, Feilden, Clegg and Bradley Architects
Bob Ghosh, Director, Glen Howells Architects, Birmingham
Don Manton, Partner, Goddard Manton Architects
Andrew Matthews, Proctor & Matthews Architects
Stephen Proctor, Proctor & Matthews Architects
Mark Reeves, Partner, Reeves Bailey Architects
Hazel Rounding, Director, Shed KM Architects, Liverpool
Bill Thomas, retired partner, Pollard Thomas and Edwards Architects

PR consultants

Robert Barlow, Senior Partner, Robert Barlow Communications
Jane Faust, Faust Public Relations
Jacqui Power, McCann Erickson Public Relations

Social housing providers

Colin Archer, Director of Development, Islington and Shoreditch Housing Association
Alan Cheetham, Service Manager for Regeneration Projects, Maritime Housing Association
Paul Dickson, Regeneration Initiatives Manager, Maritime Housing Association
Michael Fawcett, Development Officer, Islington and Shoreditch Housing Association
Martin Lippitt, Development Manager, Guinness Housing Trust

Local authority planners

John Banbow, Development Control, Liverpool City Council
Janet Dore, Conservation Officer, Cheltenham Borough Council
Nigel Edmondson, Design Section, Planning Department, Birmingham City Council.
John Finney, Conservation Officer, Planning Department, London Borough of Hillingdon, Uxbridge
Peter Jenkins, Deputy Borough Technical Services Officer, Runnymede Borough Council
Lindsay Pearson, Deputy Planning Officer, Tonbridge & West Malling District Council
David Roscoe, Manchester City Council, City Centre Development Control Team Leader
Ian Washbrook, Historic Environment and Design, South Gloucestershire Council

Media

Property Journalist, Birmingham Post and Evening Mail
Gill Burdett, Property Journalist, Manchester Evening News
Tony Chapman, Head of Awards, Royal Institute of British Architects
Hilary Clarke, Head of Communications, Commission for Architecture and the Built Environment
Sharon Martin, Property, Daily Telegraph
Janice Morley, Editor, Homes and Property Supplement, Evening Standard
Sebastian O'Kelly, Property Editor, Mail on Sunday
Martin Smith, Journalist, Homes and Property Supplement, Evening Standard
Matthew Weaver, Society, Guardian Newspaper

Appendix 4
Award Winning Schemes Used as Case Studies

Barrier Point, Silvertown, London

Location Address: Barrier Point, North Woolwich Road, Silvertown, London E16

Site Context:
The project occupies a long, narrow two-hectare site at right angles to the north bank of the River Thames, with the Royal Docks to the north. It was the first residential development south of the North Woolwich Road in a long-derelict area, but overlooking a new park focusing on the Thames Barrier. The formerly industrial, derelict site to the west is now cleared and awaiting redevelopment as a mixed use project, and further higher density development is underway to the east of the park, including a second scheme by Barratt Eastern and Goddard Manton.

Development Partners

- Developer: Barratt Southern
- Architects: Goddard Manton Partnership
- Local Planning Authority: London Borough of Newham

Development History:
Barratt acquired the contaminated but cleared riverside site in 1998. They held a limited design competition selecting the Goddard Manton Partnership as architects. The site required reclaiming to Environment Agency standards. It overlooks a recently completed nine-hectare park which was the subject of an international design competition in 1995. Barratt's 44th Docklands project followed a highly successful collaboration with Goddard Manton Partnership at Pierhead Lock at the western end of the West India Docks from which many of the design ideas were derived. The scheme was completed in two phases with the 'Seven Steps' garden apartments occupied in 2000, and the landmark tower completed in 2001.

Development Details:
Barrier Point is a development of 252 concierged apartments, yielding a density of 126 dwellings per hectare. A line of seven storey apartments over an underground car park with seven glazed escape towers provide features to relieve the western elevations and the visitor car parking screened by pergolas. On the eastern elevation six wings of apartments step down, ziggurat-like, storey by storey to ground level providing large 'sun terraces' for each flat, and six enclosed gardens for communal use set slightly above the western edge of the park. Overlooking the river the block culminates in an 18 storey, cylindrical apartment tower,

the tallest yet built by Barratt in Docklands. A glass-walled riverside restaurant will occupy the ground floor. Unit sizes vary from one bedroom to a three bedroom duplex penthouse. The whole project is executed in a 1930s International Modern style, white rendered, with masts and rails to emphasise maritime references.

Photograph: Goddard Manton Architects

Details of Awards

- Best development of the Year, National Home Builder Design Awards 2001.
- Commendation for Best Apartment Building, National Home Builder Design Awards 2002.
- What House Award 2002.
- Seven other awards making it 'the most decorated contemporary housing development' in Britain.

"...simplicity, clarity and panache...ultra-modern...crisp detailing and large stepped roof terraces...a landmark in its own right..."

Mile End Road, London

Location Address: 9-25 Mile End Road, Tower Hamlets

Site Context:
An infill scheme on the north side of the busy Mile End Road, adjacent to an ornate nineteenth century fire hall and the seventeenth century Grade 1 listed Trinity Green Almshouses. This short section of the street is boulevarded so the housing fronts onto a service road behind a narrow treed green strip.

Development Partners:

- Developer: Copthorn Homes, division of Countryside Properties
- Architects: Proctor Matthews
- Local Planning Authority: Harlow District Council

Photograph: Homes For Change

Development History
Copthorn originally envisaged a neo-Georgian approach to this mixed-use development, but the first set of architects ran into difficulties with gaining planning permission. The developers had noted Proctor and Matthews contemporary prize-winning scheme in Limehouse, and gave the architects considerable freedom to develop a scheme for the site that would gain permission.

Development Details:
The development comprises six houses and 59 flats and maisonettes, eleven different house types in all, at a density of 148 dwellings per hectare. A frontage block of three storeys of flats, above ground floor commercial capable of being divided into seven units, provides a nearly continuous street frontage largely aligned with the back of the pavement. It culminates in a rotunda which leaves a small gated entry space into an internal lane and courtyard. Here two rows of four storey flats and maisonettes face each other across a street/courtyard aligned north-south.

The majority of parking is under the flats and in the courtyard. Projecting stairways and bin lockers break up the spaces. A tree is planted at the foot of each staircase and the boundary wall at the end of the courtyard is relieved by a water feature/sculpture. The raising of the courtyard and the recessing of the top storey ensure a more domestic scale to the street, while the stairways and balconies provide semi-public spaces to be inhabited. The gardens to the rear of the terrace are only 8 metres long at best, but each unit has front patio and balcony space while the top flats make good use of external roof space.

One of the features of the scheme is the privacy it provides for residents, particularly through the use of a pierced terracotta screen on the Mile End Road frontage, and the first floor entrances, balconies and screens on the rear terraces. The use of terracotta, complemented by brown timber panelling against white stucco and yellow London stock bricks, provides a warmth that complements the use of grey steel to support the screens, stairs and balconies. The whole composition has a strong contemporary aesthetic.

Details of Awards

- Guardian Housing Award 2000.
- National Home Builder Project Award 2002.
- Completed Scheme Award and Building for Life Award, Housing Design Awards 2002.

"...a groundbreaking achievement...at 148 dwellings per hectare...the design quality of the kitchens and bathrooms...storage provision and external detailing...this consistent approach to every aspect of housing design...earned it the accolade..."

New Hall Farm, Harlow

Location Address: New Hall Farm, Harlow (parcel 1b)

Site Context:
This is a greenfield site on the edge of Harlow, the second phase of a major land release that will deliver some 2,500 homes in an urban extension to the new town. The site is immediately north of Church Langley, a 1990s urban extension showing notable limitations and the experience of which has caused the landowners to attempt much more ambitious design and development standards at New Hall.

Development Partners

- Developer: Copthorn Homes, Division of Countryside Properties
- Architects: Proctor Matthews
- Local Planning Authority: Harlow District Council

Photograph: Homes For Change

Development History

The landowners have masterplanned the development, reinforcing this with a design code, both the product of Roger Evans Urban Design. The landowners changed their approach after local dissatisfaction with the results of phase one, introducing a design/developer competition for nine subsequent phases, the second of which was won by Copthorn and Proctor Matthews. The development was largely completed by summer 2003.

Development Details:

The development comprises 85 dwellings including one and two bedroom apartments and two, three and four bedroom houses as well as three affordable units to the same specification. The scheme is designed into a masterplanned, gridded street layout of perimeter blocks, with two of the larger blocks containing mews/garage courts. A modern architectural vocabulary is adopted overlaid on a townscape-driven design framework to create an interesting streetscape with strong landmark buildings anchoring the street corners.

The dwellings are not an adaptation of previous standard floor plans; they were specifically designed for the project with an emphasis on flexible layouts to enable easy adaptation as household circumstances change, a significant contribution to 'lifetime communities'.

Details of Awards

- Project Award and Housebuilder Award, Housing Design Awards 2002.

"at last the twentieth century arrives on a Greenfield site…compact, urban and eschews any pretensions to a faux-historical approach…"

Bishops Mead, Chelmsford

Location Address: Chancellor Park, Chelmer Village, Chelmsford, Essex

Site Context:
A large green field site to the east of Chelmer Village, a 1980s extension of Chelmsford, bordered on the west by the A12, the Chelmsford by-pass.

Development Partners:

- Developer: Taywood Homes (Taylor Woodrow)
- Architects: Reeves Bailey Architects
- Local Planning Authority: Chelmsford District Council

Development History:
Taywood Homes gained a 600 dwelling, outline planning permission on this 40 hectare greenfield site named Chancellor Park with a scheme master-planned by Barton Wilmore. Halfway through the first phase of 150 homes, the local planning authority expressed their dissatisfaction with the scheme and a rethink was undertaken. Reeves Bailey Architects had been commissioned by Taywood to produce a new approach to housing design and layout that would take on board the twin ethos of 'conservation/PPG3'. Bishops Mead, a 1.8 ha site, provided the opportunity to try out the approach and the architects were instructed to achieve a 15 per cent increase in dwelling units using a modern interpretation of the local vernacular. Their proposals were a full interpretation of the Essex Design Guide, delighted the local planning authority, and proved very popular with house buyers. The architects have been retained to complete later phases of the scheme

Development Details:
As the 2002 assessors note "...one of the first greenfield entries to tackle the requirements of PPG3 with a density above the 30 (actual 33) dwellings per hectare threshold, wide frontage houses brought up to the back of pavement to form a coherent street pattern, and an intelligent use of 'drive through' access under the houses to take cars off the street without, in the main, the need for garage courts…". The estate layout and the 1 to 6 bedroom house types are a marked departure from the earlier phases of the scheme, which used standard

detached 'Vicwardian' house types. Three storey Georgian reproduction townhouses abut a variety of cottage and Essex Design Guide styles to fully enclose the street with pitched roof and traditionally gated car ports completing the illusion of a village. Informal urban spaces are created at occasional intersections fronted by the larger "Georgian" buildings and planted with street trees. A palette of traditional Essex building materials is used, but with a widened colour palette, and sash and casement windows (but in uPVC) reinforce the traditional character with a variety of vernacular details. Reduced road widths and almost non-existent pavements combine with a fine grained street grid that allows pedestrian, but not vehicle, through movement to create an intimate streetscape. This is reinforced by 'townscape' design principles with landmark buildings as focal points for street corners, junctions and public spaces and carefully closed vistas. Later phases of the scheme (4-5 by Bryant Homes) have adapted these principles to include assisted ownership flats for teachers, and 13 different house types.

Local Planning Authority Response:
The local planning authority expressed disappointment with an earlier phase of the scheme and was instrumental in Taywood rethinking their approach. They considered that the approved scheme was the best interpretation to date of the Essex Design Guide.

Photograph: Reeves Bailey Architects

Details of Awards

- Project Award, Housing Design Awards 2000.
- 'Worth a Detour'; Housing Design Awards 2002.

"...a close knit layout...density of 33 dwelling units per hectare, intelligent handling of car parking, and a very definite sense of place..."

But the Housing Design Awards 2002 citation also notes

> *"...the price of creating a new "old" Essex village is a conflict between the aspirations of owners and their willingness to pay. Look closer and the sash and casement windows are in fact uPVC...The materials, detailing and soft landscaping are no doubt low maintenance but also harsh. Perhaps this scheme is at once the first of the new, and one of the last of the old."*

Century Court, Cheltenham

Location: The scheme is located on Bath Road in Cheltenham town centre.

Site context:
The site is located opposite to Bath College (Grade 2*) which is built in a Gothic style. Farther up the road are other parts of the College that are Grade 1 and Regency. To the left of the site is a former hotel that is also Regency and Grade 2. On the right side are the backs of a Grade 2 Regency terrace, whilst the back street is Regency/Victorian transitional. Bath Road forms part of the A46 within the town centre.

Development Partners

- Developer: Beaufort Homes (now Crosby Homes Special Projects)
- Architects: Feilden, Clegg and Bradley Architects, Bath
- Local Planning Authority: Cheltenham Borough Council

Development History:
The site was previously a 1960s office block with a car park at the rear. A local architect produced a Regency pastiche scheme for the site and it was discussed with the local planning authority. It was agreed that a more assured modern scheme should be considered. Beaufort then approached Feilden, Clegg and Bradley and the built scheme was proposed based on a developer's brief that indicated the footprint, apartment mix, density and a series of desirable flat plans. The scheme was the architects' first private sector housing scheme. The £12.2 million scheme was designed in 6 weeks.

The Royal Fine Arts Commissions and English Heritage supported the scheme and so the Council was minded to approve it despite a lot of local opposition to the "modern" design. The Royal Fine Arts Commission commented that "...the architects approach to the problem of accommodating a dense residential development on a brownfield site is exemplary." They also pointed out that the architectural approach "...offers a skilful and refreshing reinterpretation of the Cheltenham terrace tradition..."

When the scheme was almost on site the developers became concerned that the scheme might not sell in Cheltenham and the company discussed redesigning the scheme with the local planning authority, but the council insisted on a modern design.

Development Details:
96 apartments and houses are provided on the 1 ha site. The scheme includes a five-storey apartment block backing onto Bath Road behind a pre-existing landscaped area. 2 four-storey drum shaped apartment blocks have then been built with access off the central courtyard. 9 town houses provide new frontage to the neighbouring street Montpellier Grove and a four-storey apartment block also backs onto this same street. Vehicular access is from the rear to underground parking. The buildings are all white rendered with sections of hardwood cladding, and outdoor amenity space is provided via balconies and roof terraces.

The Local Planning Authority Response
The Highway Engineers refused to allow either vehicular access or building frontage onto Bath Road. In addition senior offices visited the site and agreed that, contrary to Members' general aspirations, the original pastiche scheme was not appropriate given the mix of Gothic and Regency styles within the vicinity. As a result officers encouraged a contemporary design. The planners were also concerned about the building mass and its position on the Bath Road frontage: "We felt that if it came too far forward it would be overbearing and intrusive on the 2* college..." In addition they also wanted to recreate frontage on the road to the back of the scheme. The site was considered a sensitive and highly visible site, and so it was subject to a higher degree of council intervention than would normally occur for housing schemes.

Details of Awards

- "Worth a Detour", *Housing Design Awards* 2002
- "Best Apartment Building" Commendation, *National Home Builder Design Awards* 2002.

The architects put the scheme in for the *Housing Design Awards* and RIBA awards whilst the developer put the scheme in for the *National Home Builder Design Awards*. The scheme is described as"…a textbook demonstration of a contemporary high density scheme in total sympathy with its context. The site layout and provision of private open space related to different times of the day is exemplary" (*Housing Design Awards*, 2002). The *National Home Builder Design Awards* describe the scheme as "…a beacon of hope for the future of high density housing in our towns and cities…."

Tanner's Acre, Marshfield

Location Address: Tanner's Acre, Marshfield, South Gloucestershire

Site Context:
Marshfield is a traditional village on the edge of the Cotswolds that is strung out along a High Street. The Tanners Acre site is to the north west of the village centre between the High Street and the ring road, but it is connected through to the High Street with a footpath.

Development Partners

- Developer: Crest Nicholson Residential (South West) Ltd
- Architect: Crest Nicholson Residential (South West) Ltd
- Local Planning Authority: South Gloucestershire Council

Development History
The site was previously designated for housing development in the Local Plan. The site formed one part of two designated sites, and this development would have been the first phase. The second phase has subsequently been withdrawn from the plan. The site previously contained an agricultural machinery business and gardens. A development brief existed for both phases of the land produced by North Avon. Crest Nicholson submitted an application in line with the original brief for executive homes but South Gloucestershire (a new Authority) said that they were not happy with that, and requested something that a) reflected the character of the conservation area more and b) offered a higher density. This was ultimately accepted by Crest Nicholson. A new brief was produced in association with Crest Nicholson and the final scheme emerged as a result of what became a close working relationship.

Development Details:
The scheme includes 44 two, three and four bedroom houses in both detached and terraced forms in a tightly knit network of streets and lanes. The scheme includes 4 affordable houses fronting an open space with play equipment. Architecture is in a traditional idiom with stone and render finish. Particular emphasis is on townscape quality, and the scheme represents a marked improvement to the adjacent housing developed in the last 30 years. Parking is to the rear of houses on the main frontages to retain a coherent street scene.

The designers achieved an innovative form of highways contrary to local standards in order to achieve greater townscape quality. They also put street lighting on buildings to remove unsightly poles and hid meter boxes to protect the composition and quality of the facades.

Local Planning Authority Response:
The planners recognise that they presented Crest Nicholson with a challenge following their rejection of an original brief produced by the former authority. As such, they feel that they had a significant role to play in the creation of this final scheme. Many of the design characteristics that emerged resulted from their interventions through the process of developing a second brief, and they also worked hard to ensure that planning conditions were met. The Parish Council found it difficult to accept the idea that a higher density would be more sympathetic to the character of the village. They were also concerned that render rather then stone would be used on the main routes, even though many of the older buildings in the village have a similar render finish.

Most schemes in the area are not of this quality and there are virtually no schemes in a modern idiom. Some members of the Council had a particular problem with contemporary forms of design, despite officer recommendations to the contrary. Since PPG3, the authority has been committed to raising densities (50 units/hectare). They also promote brownfield development where possible, although they have limited sites. They have realised few mixed use and energy efficient schemes. They are very committed to bio-diversity and employ an ecologist to provide advice.

Details of Awards

- What House Award for the best scheme in the UK, 2001
- South Gloucestershire Design Award 2001

Britannic Park, Moseley, Birmingham

Location Address: Britannic Park, 15 Yew Tree Road, Moseley, Birmingham, B13 8NQ

Site Context:
Britannic Park is located in the residential suburb of Moseley, approximately two miles to the south of Birmingham city centre, and 15 minutes from Birmingham International Airport. The main building on the site, Britannic House, was constructed in 1962-3 as the headquarters of Britannic Assurance, within landscaped grounds.

Development Partners

- Developer: CALA Homes (Midlands) Ltd, Sheldon, Birmingham
- Architect: Malcolm Payne Design Group, Birmingham
- Local Planning Authority: Birmingham City Council

Development History
Britannic House was constructed in 1962-3 to the designs of T. M. Ashford. It was a landmark building in Birmingham, an Art deco crescent-shaped building with a creamy Travatino marble and green Westmoreland stone façade. Britannic Assurance vacated the building in 1993, and it remained vacant and vandalised until being acquired by CALA Homes (Midlands) Ltd in 1998.

Development Details:
Originally a five-storey building, CALA added an additional 6th storey to form penthouses, converted the offices into apartments, and added a new-build element comprising two 3-storey crescent-shaped buildings each with a glass atrium entrance. The development comprises 113 converted and newly-built apartments, 7 of which are penthouses and duplex penthouses. The main entrance features a stainless steel mast with canopy, and the scheme provides extensive underground parking facilities, as well as the private Britannic Club (including swimming pool, gymnasium, sauna, Jacuzzi and steam room). It is a 'gated' scheme, with extensive CCTV and concierge. The landscaped grounds are maintained communally by a management company.

Local Planning Authority Response:
The authority's attitude to housing design and to urban design in general is nationally recognised as very proactive. When Britannic Park was being developed, the City's design guidance had not been up-dated, but the City utilised the principles emerging in the new guidance in dealing with the development. The authority was involved in the scheme from the beginning, with both development control and design officers fully involved. The authority was not involved in the competition entry.

The authority considers that Britannic Park is not typical of the type of scheme typically submitted by the developers, being a residential conversion of an office building with some new-build. There has been an improvement in the quality of schemes coming forward, but this is by no means universal, and some companies have not moved forward. Companies are more open to taking design advice, due to the focus on complex sites arising from the urban renaissance agenda. The authority is promoting higher densities in appropriate locations, and this is endorsed by Council members. They are less willing to accept the crowding of dwelling units on a site favoured by some developers – a design-led approach is required, which starts with place and people rather than car-dominated layouts.

Details of Awards

- Best Brownfield/Renovation, *What House? Awards* 2002
- Joint Winner Best Restoration and Conversion, *National Home Builder Design Awards* 2002
- Best Apartment Award, *Birmingham Evening Post Design Awards* 2001

Coppermill Lock, Harefield, Middlesex

Location Address: Coppermill Lock, Harefield, Middlesex UB9 6TQ.

Site Context:
The site is located approximately 1 mile to the north-west of Harefield within the London Borough of Hillingdon. Situated on the edge of the built-up area of Harefield, adjoining the metropolitan Green Belt and adjacent to the Grand Union Canal, the River Colne and Lynsters Lake, the application site comprised approximately 21 acres. Of this, approximately 15 acres was situated in the Green Belt and remains undeveloped, the redevelopment site itself is 5.7 acres in size. A public right of way forming part of the Colne Valley Park passes through the site, which falls within the Coppermill Lock Conservation Area.

Development Partners:

- Developer: Linden Homes Chiltern Ltd, Harefield, Middlesex
- Architects: John Thompson and Partners, London
- Local Planning Authority: London Borough of Hillingdon

Development History:

The site was formerly occupied by industrial buildings (used for a mix of employment uses) and derelict structures. Having won a planning appeal with the vendor's original proposal for the site, Linden resubmitted an application in 1998 for mixed-use development (50% residential/50% employment use) and secured planning permission in five months. Development commenced in January 1999 and was completed in 2001.

Development Details:

The development comprises 42 residential units, 4 live/work, 20,000 sq ft of light industrial units, 10,000 sq ft of offices and a 40-bedroom nursing home. The scheme is designed to provide an urban design framework of linked open spaces and enclosed streets and courtyards of varying scale, creating a mixed-use cluster that maximises the canal and riverside frontage, integrating with the adjacent Green Belt. The scheme layout includes a 'village green' and a 'town square', and the housing component comprises 2 and 3 storey dwellings (detached, townhouses and semi-detached; 2-5 bedroom). Materials are vernacular brick, clay tiles, render and timber, to reflect and enhance the setting of the Coppermill Lock Conservation Area. The scheme has been carefully landscaped, and includes traffic calming to a maximum speed of 20mph. The roads are not adopted, and a management company manages common areas.

Local Planning Authority Response:

From a design perspective, the Coppermill Lock scheme was essentially straightforward. Having lost an earlier planning appeal increasing the proposed density on the site, the authority found the scheme proposed by Linden Homes and their architects, John Thompson and Partners, accorded both with the Borough's design policies and the Conservation Area designation. The authority was involved in lengthy discussions with the developer and architects from an early stage, but the scheme essentially involved only 'tweaking'. The extension of the recreational route along the canal and the treatment of the canal side itself are particularly successful elements, producing a recreational planning gain.

The attitude of the local planning authority to design is to strive to do what it can to achieve design improvements, and it is not considered that the general quality of residential development is improving greatly, particularly in the case of volume house builders. Hence schemes such as this by Linden, indicate a willingness to accept a design challenge rather than a problem.

The authority can see parts of the urban renaissance agenda becoming a problem, and it is argued that 'compact city' policies for London have to be interpreted in the context of Boroughs such as Hillingdon as some members and the public are resistant to changes in character associated with increasing densities. The authority encourages house builders to retain professional advice, but frequently volume house builders' own 'in house' architects do not produce designs which are sensitive to context. However, there is some evidence that, on large sites and, for example, hospital redevelopment schemes in the Green Belt, a more integrated design approach is emerging.

Details of Awards
'Best Brownfield Development, Joint Silver Award, What House? Awards 2002:

"...thoughtful design makes for a very pleasing development and another fine example of brownfield development".

Lacuna, West Malling, Kent

Location Address: Lacuna, Kings Hill, West Malling, Kent

Site Context:
Lacuna is the central village in the new settlement of Kings Hill, which includes a business park and golf course, that is being developed on a 263ha former airfield at West Malling, Kent.

Development Partners:
Kings Hill is a new community that is being developed by the Rouse Corporation of America in partnership with Kent County Council. Rouse Kent is developing the business park and golf course, and has sold land to almost a dozen house builders.

- Developers: Sunley Estates plc, Environ Country Homes
- Architects: Clague, Canterbury, Kent
- Local Planning Authority: Tonbridge and West Malling District Council

Development History:
Formerly a Battle of Britain airfield, Kings Hill (263 ha site) is being developed as a new settlement near West Malling by Rouse Kent. The business park will eventually employ 10,000 people in business, retail, educational, recreational and community facilities. 1,850 houses and apartments are planned, of which more than half have been built, accommodating 3,500 people (6,000 when complete). Lacuna is the name of the central village at Kings Hill.

Development Details:
The Lacuna development is planned to have 181 houses and apartments, but only the first phase (38 dwellings) had been completed at the time of the Award. All the house builders have to conform to a master plan prepared by Rouse Kent, within which they can use their own architects. Claque architects were appointed by Environ Country Homes and its joint-venture partner Sunley Estates to design the new dwellings, and were also appointed by Rouse

Kent as the masterplanner for the central area of Kings Hill, which will include shops and other facilities immediately north-west of the Lacuna housing. The brief was to create a traditional village centre in a modern idiom, and the key design principles for the Lacuna housing are to provide a mixed variety of house types and elevational treatments while respecting local architectural context and detailing. Lacuna is the only development at Kings Hill that offers modern design, albeit using traditional building materials and Kentish vernacular forms and styles, such as white weather boarding. Five of the dwellings are the 'Super E' home, an energy-efficient eco-friendly timber-framed dwelling. The layout of the houses has been designed to create safe streets, with the emphasis on pedestrian circulation and cycle ways. Communal areas are to be maintained by a management company.

Local Planning Authority Response:
Both the overall layout and the individual housing designs at Lacuna are imaginative, and not typical of the rest of the Kings Hill development. This innovative design approach was facilitated by planning officers from Tonbridge & West Malling Council, though it required a flexible attitude to both the latest Kent Design Guide and the Council's own planning standards for residential areas. Following a visit to Poundbury, members were prepared to accept the new principles of residential design exemplified there (in terms of density, layout and space), and to link this with the modern approach to housing design proposed by the developers. The result has been a radically new approach to treatment of streetscape, well removed from pastiche but acknowledging the Kent vernacular tradition.

The local planning authority's experience is that the more innovative the architect, the less tension with planning policies, since planners and architects are able to mutually identify problems and arrive at acceptable solutions. The development of Local Authority design guidance has had an impact (Kent Design Guide), as has the Government's agenda with associated increases in density. Elected representatives have sometimes been reluctant to accept the latter, as well as reduced car-parking, but in time will probably grow more accustomed to these changes.

Details of Awards
Plot 39 (the Ricard house type) won:

The *Evening Standard New Homes Awards*;
National Home Builder Design Award, 2002;
What House? Awards, 2002.

The scheme was also selected for exhibition in the Housing Design Awards, 2002.

"The judges felt that the good proportions and satisfying contemporary detailing of this high density development, which satisfies the requirements of Planning Policy Guidance Note 3 (PPG3), was worthy of the award as the Best House in England" (National Home Builder Design Award, 2002).

Virginia Park, Stroude Road, Virginia Water, Surrey

Location Address: Virginia Park, Stroude Road/Christchurch Road, Virginia Water, Surrey, GU25 4FN

Site Context:
The site includes the former Royal Holloway Sanatorium and its grounds at Virginia Water, Surrey, close to the A30 and the M25/M3 junction 2 interchange.

Development Partners:

- Developer: Octagon Developments Ltd, East Molesey, Surrey
- Architect: Howell Smith Partnership, ML Design Group, London
- Local Planning Authority: Runneymede Borough Council

Development History:
Built between 1873 and 1885 by Thomas Holloway and designed by William Crossland, the Royal Holloway Sanatorium (a 480-room building) served as a "hospital for the insane of the middle class" until 1948 when it was transferred to the National Health Service. The hospital closed in 1981, when it was sold to a property company which planned to convert it into 120,000 sq ft of offices. After changing ownership several times and ending up in the hands of a receiver, the building lay empty and vandalised for 12 years, becoming completely derelict. Octagon purchased the building in 1994, and submitted a planning application for the restoration of the buildings and their conversion into townhouses, with new houses and apartments in the grounds (213 units).

Development Details:
The scheme involved the complete restoration of the Grand Hall in the main Sanatorium building (Grade 1 Listed), the conversion of the main building to 23 large townhouses and the construction of 190 new houses and apartments designed to be in keeping with the 24 acres of grounds. The completed development has a spa complex and gymnasium, multi-purpose sports hall (in the former Chapel that is Grade 2 listed), a great hall, all-weather tennis court and is 'gated'. Octagon spent 6 years on a programme of restoration and conversion, described by English Heritage as "one of the largest art-conservation projects in Europe".

Photograph: Octagon Developments

Local Planning Authority Response:
Runneymede is highly committed to the urban design agenda, encouraging high-quality, imaginative housing schemes. Its positive attitude is reflected in the Borough's own design awards, in operation for some ten years. The Borough Council had been concerned about the future of the Virginia Park site since the closure of the hospital in 1981, with a succession of proposals for office, hotel development and residential redevelopment failing to come to fruition. The Borough co-operated with English Heritage and the Victorian Society to formulate a balance between enabling and conservation when Octagon presented their proposals for redevelopment. The Borough worked closely with Octagon over a six-year period on the site's redevelopment, with a management/supervisory team comprised of representatives of Octagon and their architects, English Heritage and the Borough Council's Technical Services Department meeting regularly on site. An excellent working relationship with Octagon was established, with a good, responsive input from the architects.

In terms of the urban renaissance agenda, the Authority initially witnessed some caution amongst some elected members with regard to potential issues of 'town cramming' and reduced car-parking, but responses have generally been good. There is a definite perception of improved quality in housing schemes over the last five years, and the Borough sees a positive role for design competitions.

Details of Awards
Housing Development of the Year, *National Home Builder Design Award* 2000 –

The development has won 24 other awards over a five-year period and the Company also enters the *Evening Standard New Homes Awards*; *What House? Awards*; *Greenleaf Awards*. The company has also been House Builder of the Year several times since 1991.

Timber Wharf, Worsley Street, Manchester

Location Address: Worsley Street, Manchester

Site context:
The site is on the western edge of Manchester city centre adjacent to the Castlefield area and bordering the Bridgewater Canal. Castlefield has seen extensive urban renewal of former warehouses into office, residential and A3 uses, and high quality landscaping has significantly enhanced the image of the area. A number of private house builders continue to develop apartments in Castlefield. Timber Wharf is located on the other side of the Manchester inner ring road from Castlefield. Its immediate context is of old industrial buildings, poor quality original public realm, and an adjacent new apartment development which would have been the first indication that the development pressure experienced in Castlefield was spilling over into this area. Part of the area falls within the Castlefield conservation area.

Development Partners

- Developer: Urban Splash
- Architects: Glen Howells Architects, Birmingham
- Local Planning Authority: Manchester City Council

Development History:
Manchester city centre has seen a rapid development of its city centre housing market, especially in sites outside of the core area, to the point where now it is regarded as the most profitable form of development in a large number of locations.

Urban Splash recognised the development potential in this area which has evolved progressively following the success of Castlefield. They completed two initial schemes in the area: The Britannia Mills and the Box Works. They continue to develop neighbouring sites in what they call Britannia Basin.

Urban Splash commissioned an RIBA design competition to select both a prototype new build scheme and also a new architectural practice to work with. The local planning authority informed the competition brief and a commitment was made by Urban Splash to build the final scheme. 434 expressions of interest and 162 entries were received. Judging was done on merit alone as scheme designers were kept anonymous.

Development Details:
The scheme forms a single apartment block that contains 181 penthouses, live/work units, and apartments. The scheme includes 9 different types of layout within the scheme. It also includes two decks of underground secure parking and a shared private garden adjacent to the canal. Public access to the canal has been retained. Live/work units are for sale at £290,000. Normal apartments are for sale at £190,000. Urban Splash retain an interest in the building as their head office is located there.

Photographs: Urban Splash

Local Planning Authority Response:

The local planning authority were invited to contribute to the competition brief. They required the design to show sympathy for the conservation area, for public access to the canal to be retained, and they wanted to ensure that the scheme became properly integrated with both neighbouring developments and the wider area.

The scheme received planning permission without controversy. There is no neighbouring residential population, English Heritage supported the proposal and the members were also happy with the scheme.

Details of Awards
Project Award, Housing Design Awards 2000

"This scheme is the outcome of an imaginative developer led competition...calling for 'clearly contemporary' homes...at costs equivalent to those achieved by volume housebuilders...the team have exploited the cost and quality advantages of fast-track construction...offering a liberating glimpse of the future of urban regeneration..."

Chillingworth Road, Lower Holloway, London

Location Address: Chillingworth Road, Mackenzie Road, and Lowther Road, Islington, London

Site Context:
This is a former college site just off the Holloway and Liverpool Roads in London. The context is a fragmenting townscape with a mix of remnant Victorian buildings and earlier cottages, terminating in a fine, neo-classical church (now in university use). To the east a 1980s health centre faces the project while to the west are a series of 1950s council flats, some facing Mackenzie Road but others set in illegible space with no regard to the road pattern.

Development Partners

- Developer: Galliard Homes
- Housing Associations: Guinness Trust and Islington and Shoreditch Housing Association
- Architects: Pollard Thomas and Edwards
- Local Planning Authority: London Borough of Islington

Development History:
The architects often act as developers and they put in a bid for the site when it was marketed. Formerly owned by the City of Islington College, the architects put together a joint venture, mixed tenure scheme that exceeded the prevailing planning requirements in Islington. This obviated the need for a S106 agreement with planning permission and as a result the project received financial support from the local housing fund. The architects brought in the Guinness

Trust and the Islington and Shoreditch Housing Association to provide the social housing and Galliard Homes as the site developers and builders. The architects have a strong commitment to mixed communities and by initiating such projects retain much more control over both design and development content. They had worked with Galliard before and the scheme appealed to the developers despite the fact that they do not normally build houses.

Development Details:
The development includes 18 2- and 3- bedroom private houses and 13 2-bedroom flats fronting Chillingworth and Lowther Roads, with parking on the frontage. The perimeter block is completed by 23 1 and 2 bedroom flats on Mackenzie Road owned by the social housing providers with 8 1- and 2- bedroom flats in shared ownership with Guinness. The 3 to 4 storey massing respects the adjoining 1950s council flats on Mackenzie Road and is modest compared with a series of 8 storey squat towers to the north and east. The streets are re-enclosed and 7 mature street trees retained, with 7 new ones planted. The interior of the block accommodates private, fenced gardens and additional individual car parking spaces for the houses as well as a communal green. The social housing has ground floor patios and a fenced play area. The project achieves a density of 150 dwellings per hectare.

A feature of the scheme is its sensitivity to the 1950s 4-storey Greater London Council scheme on Mackenzie Road adopting similar massing and materials (London stock bricks). The elevations look remarkably minimalist/modernist on plan but their site planning, materials and careful detailing ensure they recreate three traditional north London streets in an urbane and refined manner. The architects worked carefully with local residents to take on board their concerns about design, altering the site planning to protect views.

Details of Awards:

"Worth a Detour", *Housing Design Awards* 2001

The Scheme also features in the Urban White Paper 2000 as *"...a convincing demonstration that skilled designers, working closely with supportive clients and local planning officers, can produce housing of an appropriate urban scale and type for an integrated mix of tenures and household sizes..."*

Preston Point, Preston Street, Liverpool

Location Address: The scheme is located in Preston Street in the city centre of Liverpool.

Site context:
The site is located in the city core of Liverpool and it forms part of a triangular block which is in various stages of renewal. The site is within a conservation area and is also about 300 metres from Lime Street Station, falling roughly between the office, retail and museum districts. Preston Street is a quiet street running perpendicular to busier streets which form part of the city centre one way system. Adjacent to the site is a recently refurbished Old Haymarket mixed use scheme by Urban Splash and Maritime Housing, a newly constructed Travel Lodge Hotel by Shed KM, and a development site for a proposed commercial building also designed by Shed KM for Urban Splash. The scheme itself is a refurbishment of an old warehouse building.

Development Partners

- Developer: Maritime Housing Association, Liverpool
- Architects: Shed KM, Liverpool
- Local Planning Authority: Liverpool City Council

Development History:
Preston Point was a derelict warehouse building located on a very tight site adjacent to recent developments. Urban Splash suggested that Maritime Housing consider both converting the building for housing as part of its "Living in the City" initiative, and also that they use Shed KM as architects following their design and commercial success with the nearby "Collegiate" development.

The scheme required that architects design 14 apartments into the scheme. Working with Urban Splash, who owned the neighbouring site, allowed them to put windows in a blank façade which made the development viable.

Development Details:
There are 15 shared ownership apartments along with ground floor commercial uses. There is no car parking, but the scheme was required to include cycle parking spaces. Apartments have balconies.

The shared ownership involves 75% – 80% shared equity with the rest of the equity being retained by Maritime Housing. Sale prices varied from £52,500 – £60,000. A rent/service charge of £20 per month is also charged.

Local Planning Authority Response
Prior to the development the planning authority considered the building for listing but this was rejected. The City Council required that the proposed balconies be shielded to limit overlooking from the proposed neighbouring development and also that cycle parking be included. Discussions with the City's conservation officer required that particular window forms be used in the Preston Street façade where the original façade was being retained.

The application was regarded as not controversial.

Details of Awards

- Project Award and Regeneration Award, *Housing Design Awards* 2000
- Completed Scheme Award, *Housing Design Awards* 2002
- *Civic Trust Award*

The scheme was entered by the architects in agreement with the client. The scheme won the *Housing Design Awards* because it was loft development by a housing association, and because the design was regarded as a creative conversion on a very limited site and with a very limited budget. Of particular merit were the successful balconies, the exploitation of the original vaulted ceilings, the creative internal planning of the apartments, and the "*...meticulous attention to planning and detail.*"

Bibliography

Adams & Watkins, (2002) Greenfields, Brownfields and Housing Development, Oxford: Blackwell

Alcock, A.. (1993), Aesthetics and Urban Design, in R. Haywood and S. McGlynn (Eds.), Making Better Places – Urban Design Now, London, Butterworths.

Anonymous (2002) Home: a place to live. The Housing Design Awards 2002, unpublished awards report from the Housing Design Awards Office, Birmingham

Anonymous (2002a) *National Home Builder Design Awards* 2002: Judges Report

Bentley, I. (1999) Urban Transformations: Power, People and Urban Design, London: Routledge

Biddulph, M. (2001) Home Zones: a planning and design handbook, Bristol: Policy Press

Bourassa, S.C. (1991) The Aesthetics of Landscape, London, Belhaven Press.

Bristol City Council (1998) Living in the City: City centre housing design guide, Bristol: BCC

Brower, S. (1998) Design in familiar places: what makes home environments look good, New York, Praeger.

CABE (2002) Bungalows are still people's choice, according to MORI poll, CABE Press Release, 25 June

CABE (2003) The value of housing design and layout, London: Thomas Telford

Carmona, M. Punter J.V. and Chapman, D. (2002) From design policy to design quality, London: Thomas Telford

Cooper Marcus, C. and Sarkissian, W. (1986) Housing as if people mattered: site design guidelines for medium-density family housing, Berkeley: University of California Press

Cuff, D. (1991), Architecture – the culture of practice, Cambridge, Ma., MIT Press.

Devlin, K. (1990), An examination of architectural interpretation: architects versus non-architects, Journal of Architecture and Planning Research 7(3), 235-244.

DoE (1980), Circular 22/80: General Policy and Principles, London, DoE.

DoE/DoT (1992) Design Bulletin 32, Residential Roads and Footpaths Layout Considerations, Second Edition, London: HMSO

DoE/DoNH (1994) PPG15: Planning and the Historic Environment, London: HMSO

DETR (1998) Places, Streets and Movement: A companion guide to Design Bulletin 32: Residential Roads and Footpaths, London: DETR

DETR (2000) PPG3: Housing, London: HMSO

DTLR (2000) By Design: Urban Design in the Planning System: Towards better practice, London: Thomas Telford

DTLR (2001) PPG1: General Policy and Principles, London: HMSO

DTLR/CABE (2001) By Design: Better Places to Live: A companion guide to PPG3, London, Thomas Telford

Edwards, A.T. (1944), Good and Bad Manners in Architecture, London, Philip Allen

Essex County Council (1973), A Design Guide for Residential Areas, Chelmsford, Essex County Council

Essex Planning Officers' Association (1997) The Essex Design Guide for Residential and Mixed Use Areas, Chelmsford: EPOA

Groat, L.N. (1994), Carbuncles, Columns and Pyramids: Lay and Expert Evaluations of Contextual Design Strategies, in B.C. Scheer and W.F.E. Preiser (Eds.), Design Review: Challenging Urban Aesthetic Control, New York, Chapman and Hall, 156-164

Hooper & Nicol, (1999) The design and planning of residential development: standard house types in the speculative house building industry, in Environment and Planning B: Planning and Design, Vol. 26, pp 793-805

Hooper & Nicol, (2000) Design practice and volume production in speculative house building, construction management and economics, Vol. 18, pp 295-310

House Builders Federation (2002) Pick of the Crop – Housing Design Awards 2002

Housing Forum (2001) National Customer Satisfaction Survey, London: The Housing Forum (www.thehousingforum.org.uk)

Hubbard, P. (1994), Professional vs. Lay Tastes in Design Control – an empirical investigation, Planning Practice and Research 9(3), 271-287

Institute of Highway Incorporated Engineers (2002) Home Zone Design Guidelines, London: IHIE

Kent Planning Officers Association (2000) Kent Design: a guide to sustainable development, Maidstone: Kent County Council

Knox, P.L. and Cullen, J. (1981), Planners as Urban Managers – an examination of the attitudes and self image of senior British planners, Environment and Planning A 13, 885-898

Llewelyn-Davies (2000) Urban Design Compendium, London: English Partnerships/ Housing Corporation

Manser, M. (1983), Barriers to Design, Journal of the Royal Institute of British Architects 86, 401-403

Moro, P. (1958), Elevational Control, Architects' Journal 127, 203

Mulholland Research Associates (1995), Brand New Home Owner Attitude Survey, London, NHMB/HBF

Nasar, J.J. (1994), Urban Design Qualities: The Evaluative Qualities of Building Exteriors, Environment and Behavior, 26(3), 377-401

Nasar, J.L. (1989), Symbolic Meaning of House Styles, Environment and Behavior 21(3), 235-257

Nasar, J.L. (2000), Design by Competition: Making Design Competition Work, Cambridge, Cambridge University Press

Nasar, J.L. and Grannis, P. (1999), Design Review Reviewed: Administrative versus Discretionary Methods, Journal of the American Planning Association 65(4), 424-433

New Homes Marketing Board (2002) NHMB/Halifax Customer Survey, www.new-homes.co.uk/customer_survey/key_findings.html

Nicol & Hooper (1999) Contemporary change and the house building industry: concentration and standardisation in production, in Housing Studies, Vol. 14(1), pp 57-76

ODPM (2003) Sustainable Communities: Building for the Future, London: ODPM

Popular Housing Forum (1997), Kerb Appeal: the external appearance and site layout of new housing, Winchester, PHF

Punter, J.V. (1994), Aesthetics in Planning, in H. Thomas (Ed), Values and Planning, London, Avebury, 38-67

Punter, J.V. and Carmona, M. (1997) The Design Dimension of Planning: Theory, Content and Best Practice for Design Policies, London: E and F N Spon

Royal Fine Art Commission (1994), What Makes a Good Building? London, RFAC

Salaman, G. (1974), Community and Occupation: an exploration of work/leisure relations, Cambridge, Cambridge University Press

Stamps, A.E. and Nasar, J.L. (1997), Design Review and Public Preferences: Effects of geographic location, public consensus, sensation seeking, and architectural styles, Journal of Environmental Psychology 17, 11-32

Urban Task Force (1999) Towards an Urban Renaissance, London: DETR

Uzzell, D. and Leward, K. (1989), The Psychology of Landscape, Landscape Design 180, 34-5

Vischer J. C. and Cooper Marcus C. (1986) Evaluating Evaluation: Analysis of a Housing Design Awards Program, in Places, Vol. 3(1), pp 66-85